Forschungsberichte aus dem Lehrstuhl für Regelungssysteme

Technische Universität Kaiserslautern

Band 16

Forschungsberichte aus dem Lehrstuhl für Regelungssysteme

Technische Universität Kaiserslautern

Band 16

Herausgeber:

Prof. Dr. Steven Liu

Markus Bell

Optimized Operation of Low Voltage Grids using Distributed Control

Logos Verlag Berlin

λογος

Forschungsberichte aus dem Lehrstuhl für Regelungssysteme
Technische Universität Kaiserslautern

Herausgegeben von
Univ.-Prof. Dr.-Ing. Steven Liu
Lehrstuhl für Regelungssysteme
Technische Universität Kaiserslautern
Erwin-Schrödinger-Str. 12/332
D-67663 Kaiserslautern
E-Mail: sliu@eit.uni-kl.de

Bibliographic information published by the Deutsche Nationalbibliothek

The Deutsche Nationalbibliothek lists this publication in the Deutsche Nationalbibliografie; detailed bibliographic data are available on the Internet at http://dnb.d-nb.de .

ISBN 978-3-8325-4983-1
ISSN 2190-7897

Logos Verlag Berlin GmbH
Comeniushof, Gubener Str. 47,
10243 Berlin
Tel.: +49 (0)30 / 42 85 10 90
Fax: +49 (0)30 / 42 85 10 92
http://www.logos-verlag.de

Optimized Operation of Low Voltage Grids using Distributed Control

Optimierter Betrieb von Niederspannungsnetzen mittels verteilter Regelungsmethoden

Vom Fachbereich Elektrotechnik und Informationstechnik

der Technischen Universität Kaiserslautern

zur Verleihung des akademischen Grades

Doktor der Ingenieurwissenschaften (Dr.-Ing.)

genehmigte Dissertation

von

M. Sc. Markus Bell

geboren in Tauberbischofsheim

D 386

Tag der mündlichen Prüfung:	02.11.2018
Dekan des Fachbereichs:	Prof. Dr.-Ing. Ralph Urbansky
Vorsitzender der Prüfungskommission:	Prof. Dr.-Ing. Norbert Wehn
1. Berichterstatter:	Prof. Dr.-Ing. Steven Liu
2. Berichterstatter:	Prof. Dr.-Ing. Dirk Westermann

Acknowledgements

First of all, I would like to express my gratitude to Prof. Dr.-Ing. Steven Liu for giving me the possibility to pursue a Ph.D. work under his supervision. I very much appreciated his support for my demonstration projects and my application-oriented approach that shaped my research.

Special thanks go to Prof. Dr.-Ing. Dirk Westermann for being the co-examiner of my thesis and his interest in my work.

Furthermore, I very much appreciated the various discussions with Dr.-Ing. Felix Berkel that motivated me to always reflect on my approaches. Through these discussions, I ended up with a more profound understanding, even though we had a very different background. You have a great part in the success of this work. Thank you for the collaboration in several papers.

During my time at the institute, two students also provided a lot of input to this work. Simon Fuchs with his diploma thesis, which gave me a lot of new insides into distributed control of power grids. And Dominik Sokoluk who assisted me with his hands-on skills in the laboratory. Both of you have a great part in the success of this work through your long time of support. I am very thankful for your collaboration.

The "Nachwuchsring" institution of the University of Kaiserslautern provided me with a one-year scholarship in 2017. I'm very grateful for their financial support, as it helped me to focus more on my research during the last few steps of the journey.

I also want to thank my colleague Sebastian Caba with whom is spent many long days working in the tin can that was our office. He introduced me to the world of true sarcasm, and I will always be thankful for that. I also would like to thank my project mentor, Dr.-Ing. Tim Nagel, for giving me all the tools I needed to be successful at managing research projects. To our technicians at the institute, I would like to express my gratitude for their professional leadership and their guidance in the arena. Thanks to all the past and present colleagues at the Institute: Dr.-Ing. Fabian Kennel, Dr.-Ing. Sven Reimann, Dr.-Ing. Jianfei Wang, Dr.-Ing. Wei Wu, Dr.-Ing. Sanad Al-Areqi, Markus Lepper, Jawad Ismail, Alen Turnwald, Swen Becker, Thomas Janz, Peter Müller, Jun.-Prof. Dr.-Ing. Daniel Görges, Tim Steiner, Yun Wan, Yakun Zhou, Hengjy Wang, Min Wu, Yanhao He, Filipe Figueiredo for the good times.

Last but not least, there is my family, which I thank for endless support and their understanding in this intensive period of my life. You have been there for me, even though mostly I couldn't. Above everyone, special thanks to my wife Anika and my son Johann, who gave me the strength to endure every difficult situation. I also like to thank my brother Matthias Bell, for taking his time proofreading. This thesis could not have been completed without all of you.

From a technical point of view, nothing beyond books, real people descended from primates, and my inefficient little brain gave me anything as an input for this thesis. So closing the preface with some quotes from an intelligent and underrated man (Christopher Hitchens) that opened my eyes to pacifism and intelligent critique on religion seems appropriate.

"*Religious ideas, supposedly private matters between man and god, are in practice always political ideas.*"

"*Owners of dogs will have noticed that, if you provide them with food and water and shelter and affection, they will think you are a god. Whereas owners of cats are compelled to realize that, if you provide them with food and water and shelter and affection, they draw the conclusion that they are gods.*"

Mannheim, August 2019

Markus Bell

Contents

List of Figures

List of Tables

Acronyms

AC	Alternating current
ADMM	Alternate direction method of multipliers
CIGRÉ	Conseil International des Grands Réseaux Électriques
CLS	Controllable local systems network of the smart-meter-gateway
DC	Direct current
DCM	Direct current machine
DG	Distributed generation
DistFlow	Distribution power flow
DMPC	Distributed model predictive control
DSO	Distribution system operator
DOPF	Dynamic optimal power flow
EMA	External market authority
EN	European standard
EV	Electric vehicle
ExtLinDistFlow	Extended linear distribution power flow
GWA	Gateway administrator
HAN	Home area network of the smart-meter-gateway
HV	High voltage
LCPF	Linear coupled power flow
LIPF	Linear interpolation power flow
LinDistFlow	Linear distribution power flow
LMN	Local metro-logical network of the smart-meter-gateway
LV	Low voltage
MPC	Model predictive control
MV	Medium voltage
OPF	Optimal power flow
PMU	Phase measurement unit
PLC	Power line communication
PV	Photo-voltaic
PVC	Poly vinyl chloride
QPQC	Quadratically constraint quadratic program
RPS	Rapid prototyping system
SDP	Semi definite program
SM	Smart-meter

SMGW	Smart-meter-gateway
SYM	Synchronous machine
TOPF	Three-phase optimal power flow
VDE	German Commission for Electr. and Information Technologies

Symbols and notation

General notation

The following notation is used throughout the thesis: A complex scalar variable is denoted with $\underline{a}$ where complex vectors are bold as $\underline{\boldsymbol{a}}$ and matrices are in capital letters $\underline{\boldsymbol{A}}$ respectively. Variables without an underline e.g. $\boldsymbol{a}$ and $\boldsymbol{A}$ denote real valued vectors and matrices. $\underline{\boldsymbol{a}}^*$ denotes the conjugate complex of $\underline{\boldsymbol{a}}$ for scalars, vectors and matrices. The conjugate complex matrix $\underline{\boldsymbol{A}}^*$ is the element-wise conjugation of each entry of $\underline{\boldsymbol{A}}$. The component wise real part of a complex matrix is denoted as $\Re\{\underline{\boldsymbol{A}}\}$ and the imaginary part $\Im\{\underline{\boldsymbol{A}}\}$ respectively. As later on both single-phase and three-phase grid models are investigated, the before introduced notation extends for vectors as calligraphic $\underline{\boldsymbol{\mathcal{A}}}$ and matrices as blackboard bold $\underline{\mathbb{A}}$ for both complex and real valued. Diagonal matrices are denoted as $\boldsymbol{A} = \mathrm{diag}\left(a_1, a_2, \cdots, a_N\right)$ and row vectors as $\boldsymbol{a}^{\mathrm{T}} = \left[a_1, a_2, \cdots, a_N\right]^{\mathrm{T}}$. If values are measured from the power grid they are denoted with an over-line as $\overline{\underline{\boldsymbol{a}}}$, $\overline{\underline{\boldsymbol{A}}}$, $\overline{\underline{\mathcal{A}}}$, $\overline{\underline{\mathbb{A}}}$.

Furthermore, the euclidean-norm is defined as $\|\boldsymbol{a}\| = \sqrt{\boldsymbol{a}^{\mathrm{T}}\boldsymbol{a}} = \sqrt{a_1^2 +, \cdots, + a_N^2}$. The Kronecker product $\boldsymbol{C} = \boldsymbol{A} \otimes \boldsymbol{B}$ is defined as $\boldsymbol{C} = (\boldsymbol{a}_{ji} \cdot \boldsymbol{B})$ with j the rows and i the columns of matrix $\boldsymbol{A}$. The inverse of a matrix is denoted as $\boldsymbol{A}^{-1}$ defined such that the multiplication $\boldsymbol{A}^{-1}\boldsymbol{A} = \boldsymbol{I}$, where $\boldsymbol{I}$ is the unity matrix with appropriate size.

Sets

$\mathcal{M}_i$	set of nodes in sub-grid i that are operated by a controller, $\mathcal{M}_i \subset \mathcal{N}_i$
$\mathcal{N}$	set of all nodes in the grid
$\mathcal{N}_i$	set of nodes in sub-grid i, $\mathcal{N}_i \subset \mathcal{N}$
$\mathcal{O}$	set of all branches of the grid
$\mathcal{P}$	set of all time integer values in the prediction horizon
$\mathcal{S}^{[\nu]}$	local quadratic constraint set of node $[\nu]$
$\mathcal{S}^{[\nu]}_{\mathrm{loc},k+l}$	quadratic constraint set of local inverter power for controller $[\nu]$
$\mathcal{S}^{[\nu]}_{\mathrm{glo},k+l}$	quadratic constraint set of global nodal power for controller $[\nu]$
$\mathbb{S}^{[\nu]}_{\mathrm{loc},k}$	quadratic constraint set of local nodal power for controller $[\nu]$ for the balancing

Main symbols of single-phase quantities

$\overline{\underline{\boldsymbol{E}}}$ complex matrix of measured $\underline{\epsilon}_\nu$ for all nodes in the grid
$\overline{\boldsymbol{E}}^{\mathrm{Re}}$ real part of $\overline{\underline{\boldsymbol{E}}}$
$\overline{\boldsymbol{E}}^{\mathrm{Im}}$ imaginary part of $\overline{\underline{\boldsymbol{E}}}$
$\underline{\boldsymbol{E}}_0$ complex matrix of $\underline{\epsilon}_0$ of the slack bus node
$\boldsymbol{E}_0^{\mathrm{Re}}$ real part of $\underline{\boldsymbol{E}}_0$
$\boldsymbol{E}_0^{\mathrm{Im}}$ imaginary part of $\underline{\boldsymbol{E}}_0$
$\underline{\epsilon}_0$ complex vector of the inverse slack bus voltage
$\overline{\underline{\epsilon}}_\nu$ complex measured inverse voltage of node ν
$\underline{i}_\nu$ complex current of node ν
$\underline{i}_{\nu,\mu}$ complex branch current between node ν and μ
$\underline{\boldsymbol{i}}$ complex vector of stacked nodal current of all nodes in the grid
$\boldsymbol{i}^{\mathrm{Re}}$ real part of $\underline{\boldsymbol{i}}$
$\boldsymbol{i}^{\mathrm{Im}}$ imaginary part of $\underline{\boldsymbol{i}}$
$\boldsymbol{p}$ real part of $\underline{\boldsymbol{s}}$ (active power)
$\boldsymbol{q}$ imaginary part of $\underline{\boldsymbol{s}}$ (reactive power)
$\boldsymbol{R}$ real part of $\underline{\boldsymbol{Z}}$, matrix of grid resistances
$\underline{s}_\nu$ complex apparent power of node ν
$\underline{\boldsymbol{s}}$ complex vector of stacked apparent power of all nodes in the grid
$\overline{\underline{s}}_\nu$ complex measured apparent power of node ν
$\overline{\underline{\boldsymbol{S}}}$ complex matrix of measured $\underline{s}_\nu$ for all nodes in the grid
$\underline{u}_\nu$ complex voltage of node ν
$\underline{u}_0$ complex vector of the slack bus voltage
$\underline{\boldsymbol{u}}$ complex vector of stacked nodal voltage of all nodes in the grid
$\boldsymbol{u}^{\mathrm{Re}}$ real part of $\underline{\boldsymbol{u}}$
$\boldsymbol{u}^{\mathrm{Im}}$ imaginary part of $\underline{\boldsymbol{u}}$
$\boldsymbol{X}$ imaginary part of $\underline{\boldsymbol{Z}}$, matrix of grid reactance
$\underline{\boldsymbol{Z}}$ complex vector of stacked nodal voltage of all nodes in the grid

Main symbols of multi-phase quantities

$\mathbb{B}_{\mathrm{sh}}$ imaginary part of the inverse shunt impedance matrix $\underline{\mathbb{Z}}_{\mathrm{sh}}^{-1}$
$\overline{\underline{\mathbb{E}}}$ stacked measured complex valued matrix with $\underline{\epsilon}_\nu^{\mathrm{a,b,c}}$ for all nodes and phases
$\overline{\mathbb{E}}^{\mathrm{Re}}$ real part of $\overline{\underline{\mathbb{E}}}$
$\overline{\mathbb{E}}^{\mathrm{Im}}$ imaginary part of $\overline{\underline{\mathbb{E}}}$
$\underline{\mathbb{E}}_0$ stacked complex valued matrix with $\underline{\epsilon}_0^{\mathrm{a,b,c}}$ for the slack node and all phases
$\overline{\mathbb{E}}_0^{\mathrm{Re}}$ real part of $\overline{\underline{\mathbb{E}}}$
$\overline{\mathbb{E}}_0^{\mathrm{Im}}$ imaginary part of $\overline{\underline{\mathbb{E}}}$

$\mathbb{G}_{\text{sh}}$	real part of the inverse shunt impedance matrix $\underline{\mathbb{Z}}_{\text{sh}}^{-1}$
$\underline{\boldsymbol{\mathcal{I}}}$	vector of complex valued nodal currents stacked for all nodes and all phases
$\boldsymbol{\mathcal{I}}$	real valued vector of nodal currents stacked for all nodes and all phases
$\boldsymbol{\mathcal{I}}^{\text{Re}}$	real part of $\underline{\boldsymbol{\mathcal{I}}}$
$\boldsymbol{\mathcal{I}}^{\text{Im}}$	imaginary part of $\underline{\boldsymbol{\mathcal{I}}}$
$\underline{i}_{\nu}^{\text{a}}$	complex current of node ν and phase a
$\underline{i}_{\nu}^{\text{b}}$	complex current of node ν and phase b
$\underline{i}_{\nu}^{\text{c}}$	complex current of node ν and phase c
$\underline{i}_{\nu}^{\text{n}}$	complex current of the neutral conductor of node ν
$\boldsymbol{\mathcal{P}}$	real part of $\underline{\boldsymbol{\mathcal{S}}}$, three-phase active power vector
$\overline{\mathbb{P}}$	real part of $\overline{\mathbb{S}}$, three-phase active power matrix
$\boldsymbol{\mathcal{Q}}$	imaginary part of $\underline{\boldsymbol{\mathcal{S}}}$, three-phase reactive power vector
$\overline{\mathbb{Q}}$	imaginary part of $\overline{\mathbb{S}}$, three-phase reactive power matrix
$r_{\nu,\mu}^{\text{n}}$	real part of $\underline{z}_{\nu,\mu}^{\text{n}}$, branch resistance
$\mathbb{R}$	real part of the impedance matrix $\underline{\mathbb{Z}}_{\text{abcn}}$
$\underline{s}_{\nu}^{\text{a}}$	complex apparent power of node ν and phase a
$\underline{s}_{\nu}^{\text{b}}$	complex apparent power of node ν and phase b
$\underline{s}_{\nu}^{\text{c}}$	complex apparent power of node ν and phase c
$\underline{\boldsymbol{\mathcal{S}}}$	vector of complex valued nodal apparent power stacked for all nodes and all phases
$\boldsymbol{\mathcal{S}}$	real valued vector of nodal active an reactive power $\boldsymbol{\mathcal{S}} = \left[\boldsymbol{\mathcal{P}}^{\text{T}}\ \boldsymbol{\mathcal{Q}}^{\text{T}}\right]^{\text{T}}$
$\boldsymbol{\mathcal{S}}^{[\mathcal{N}_i]}$	real valued vector of nodal active and reactive power for grid segment $\mathcal{N}_i$
$\overline{\mathbb{S}}$	stacked measured complex valued matrix with $\underline{s}_{\nu}^{\text{a,b,c,n}}$ for all nodes and phases
$\underline{u}_{\nu}^{\text{a}}$	complex voltage of node ν and phase a
$\underline{u}_{\nu}^{\text{b}}$	complex voltage of node ν and phase b
$\underline{u}_{\nu}^{\text{c}}$	complex voltage of node ν and phase c
$\underline{u}_{\nu}^{\text{n}}$	complex voltage of the neutral conductor of node ν
$\underline{u}_{\nu}^{\text{pos}}$	complex positive sequence voltage of node ν
$\underline{u}_{\nu}^{\text{neg}}$	complex negative sequence voltage of node ν
$\underline{u}_{\nu}^{\text{zero}}$	complex zero sequence voltage of node ν
$\underline{\boldsymbol{\mathcal{U}}}$	vector of complex valued nodal voltages stacked for all nodes and all phases
$\boldsymbol{\mathcal{U}}^{\text{Re}}$	real part of $\underline{\boldsymbol{\mathcal{U}}}$
$\boldsymbol{\mathcal{U}}^{\text{Im}}$	imaginary part of $\underline{\boldsymbol{\mathcal{U}}}$
$\underline{\boldsymbol{\mathcal{U}}}_0$	vector of complex valued slack bus voltages for all phases
$\boldsymbol{\mathcal{U}}_0^{\text{Re}}$	real part of $\underline{\boldsymbol{\mathcal{U}}}_0$
$\boldsymbol{\mathcal{U}}_0^{\text{Im}}$	imaginary part of $\underline{\boldsymbol{\mathcal{U}}}_0$
$\underline{\boldsymbol{\mathcal{U}}}_{\text{L0}}$	vector of complex voltage from the no load solution for all phases
$\overline{\mathbb{U}}_0^{\text{Re}}$	inverse matrix of $\overline{\mathbb{E}}_0^{\text{Re}}$
$\overline{\mathbb{U}}_0^{\text{Im}}$	inverse matrix of $\overline{\mathbb{E}}_0^{\text{Im}}$
$x_{\nu,\mu}^{\text{n}}$	imaginary part of $\underline{z}_{\nu,\mu}^{\text{n}}$
$\mathbb{X}$	imaginary part of the impedance matrix $\underline{\mathbb{Z}}_{\text{abcn}}$

$\underline{z}^{\mathrm{a}}_{\nu,\mu}$	line impedance of the branch between node ν and μ in phase a
$\underline{z}^{\mathrm{b}}_{\nu,\mu}$	line impedance of the branch between node ν and μ in phase b
$\underline{z}^{\mathrm{c}}_{\nu,\mu}$	line impedance of the branch between node ν and μ in phase c
$\underline{z}^{\mathrm{n}}_{\nu,\mu}$	neutral wire impedance of the branch between node ν and μ
$\underline{\mathbb{Z}}_{\mathrm{abc}}$	three-phase impedance matrix of the grid without the neutral wire
$\underline{\mathbb{Z}}_{\mathrm{abcn}}$	three-phase four-wire impedance matrix of the grid
$\underline{\mathbb{Z}}_{\mathrm{sh}}$	diagonal matrix of shunt impedances

1 Introduction

1.1 Background and motivation

The German electricity infrastructure was build in a long process which started in the second half of the 19th century. At the beginning of the 20th-century electricity was already mostly produced in fossil-fueled power plants and distributed through a hierarchical structure of different grid levels [Leu09]. The transport of power within this grid structure was directed from the generation units in high-voltage levels down to end consumers on the low-voltage side. This approach was left untouched for almost a century until scientist discovered a relationship between the greenhouse gases – mostly carbon dioxide – emitted from burning fossil fuels and a change in our earth's climate [LTP07].

Today, most governments have accepted the results of climate research and the impact that fossil fuels have on the earth's ecosystem. Consequently, they started the process to reduce the carbon emission footprint of our energy production cycle. This process resulted in the progress of investments in the development of smaller and cleaner generation technologies. These so-called distributed generation units (DG), use a renew-

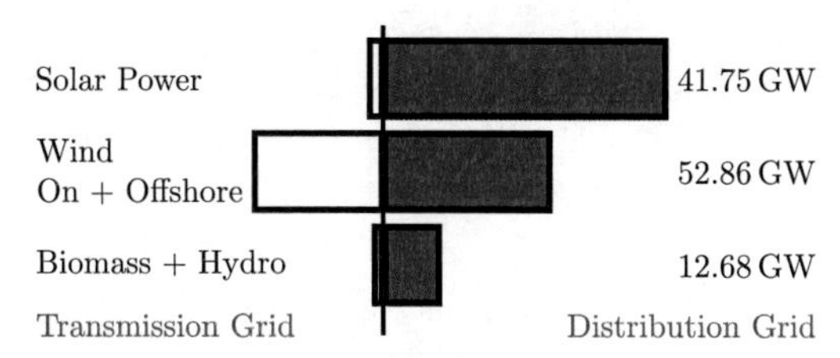

Figure 1.1: Installed capacity of distributed renewable generation units in German transmission and distribution grids [Fra17, CBCZ17]

able source of energy, e.g. wind, sun, water, which is then transformed into electricity to power the grid. The majority of DG units is connected to the distribution grids in medium- and low voltage (LV) levels, as depicted in Figure 1.1. Meanwhile numerous photo-voltaic (PV) plants account for installations in the LV grid.

Nevertheless, the electrical power system is one of the most complex installations that have ever been built by humankind. A transition from fossil energy to renewable energy sources impacts every stage in power grid operation and planning.

1.1.1 Distribution grids with intermittent generation

The expansion of renewable generation in distribution grids in the last ten years was mainly driven by monetary incentives via in-feed tariffs from the German government [Wir18]. Because of these incentives, especially for photo-voltaic generation units, the installed capacity in the German power grid has accumulated to 43.02 GW in 2017 [Qua18] and is still further increasing.

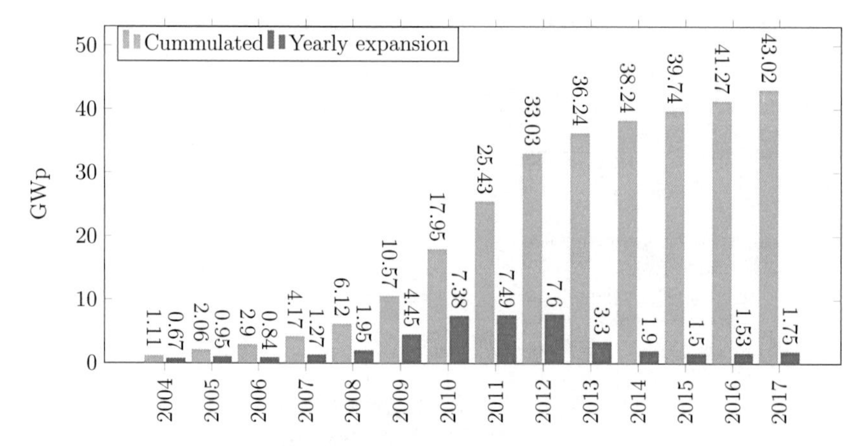

Figure 1.2: Yearly growth of installed PV capacity in gigawatt peak (GWp) in Germany [Qua18]

While the installed capacity grew very fast in the last ten years, the cost of solar modules decreased and currently hovers around 1.5 Euro/Wp [Wir18]. This is why even though the overall PV expansion in Germany currently stagnates, as depicted in Figure 1.2, the growth of installed capacity in the segment of $0 \leq P_{\text{rated}} \leq 40\,\text{kWp}$, is still stable [BNA18].

Currently, low voltage grids face several problems related to the installation of distributed generation. During low load situations, a power imbalance between generated and consumed power can be observed. The surplus of active power induces a reversing of the power flow and hence a voltage rise in the network. However, low voltage grids have traditionally not been designed to cope with generation units and as such, no mechanism is in place to operate them like medium or high voltage grids [Ker10]. In the current industrial practice several methods have been discussed or already implemented to deal with problems related to grid operation and DG:

- Upgrading power lines.
- On-line tap changing transformers.
- Reduction of active power.
- Reactive power management.

Upgrading the cables in distribution grids

A classical approach to grantee a safe operation would be to increase the LV grid capacity with additional underground cables. This way the voltage rise due to reverse power flows would be mitigated. According to a recent study [BMTM18] on short and medium term applicability this is still the favored approach by German distribution system operators (DSOs). But at best this will only postpone the system limitation and will again be an issue that operators have to face in the future. Furthermore, the approach is very conservative and expensive, as research projects that implemented intelligent solutions concluded. Only a small degree of the gained capacity through additional cables is utilized over the period of one year [BW116].

On-line tap changing transformers

To directly influence the nodal voltage in the grid, transformers with on-line tap changers (OLTC) have to be deployed. These transformers can change their turn ratio from the medium voltage (MV) to the LV side in order to reduce or increase their secondary voltage. From high voltage (HV) transmission to MV distribution grids, today most transformers are already equipped with such a tap changer [Sch15b]. The reference value for the tap changer is set remotely by the transmission system operator (TSO) [Gao13]. Some authors have already investigated the benefits of OLTC substation transformers in LV grids and possible scenarios with a local generation of the reference tap position [Oer14, LO16]. As the installation of this technology has a high investment cost, most DSOs are currently not integrating OLTCs in their grids [BMTM18].

Reduction of active power

The reduction or increase of active power at the consumer nodes has a strong influence on the nodal voltage. Hence, active power curtailment in PV systems can be used to prevent the nodal voltages from rising. Current practice to reduce the in-feed power is according to VDE-AR-N-4105 [VDE18] and limited to a stepwise reduction via a signal from the DSO. Although active curtailment is clearly not the favored method of the domestic operator to reduce voltages in the grid, it is nevertheless very effective due to the large R/X ratios of low voltage cables. Thus, it is expected, that the growing number of PV units which are installed will suffer increasingly from curtailment by DSOs to comply with capacity limitations [Sch17].

Reactive power management

In order to indirectly influence the nodal voltage, reactive power can be supplied to the grid by the inverters of DG. The standard VDE-AR-N-4105 [VDE18] states that PV inverters with a rated power between $3.68\,\text{kVA} \leq S_{\text{Rated}} \leq 13.8\,\text{kVA}$ must be able to provide reactive power support both under- and over-excited with $0.95_{\text{ind}} \leq \cos(\varphi) \leq 0.95_{\text{cap}}$. Plants with a rated power $13.8\,\text{kVA} \leq S_{\text{Rated}}$ must be able to provide reactive power in a range of $0.9_{\text{ind}} \leq \cos(\varphi) \leq 0.9_{\text{cap}}$.

Three local control mechanisms are admissible; constant $\cos(\varphi)$, variable $\cos(\varphi)$ and reactive power in relation to nodal voltage $q(\Delta \|u\|)$. The latter two are depicted in Figure 1.3.

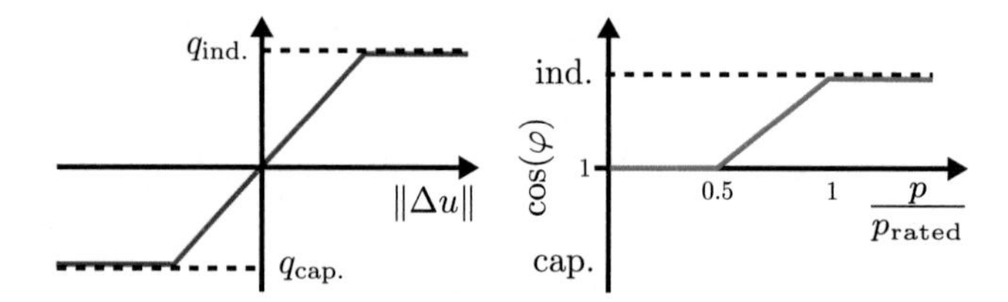

Figure 1.3: Possible local inverter profiles according to VDE-AR-N-4105

The reactive power profile must be either in accordance with VDE-AR-N-4105 or be provided by a DSO.

1.1.2 Electric vehicles and changing demand in distribution grids

Electrification of the private transportation sector shifts the need for gasoline in combustion engines to electricity stored in batteries powering electric vehicles (EVs) [Sha14, Mos16]. Motor fuels as depicted in Figure 1.4 on the left are slowly transformed into

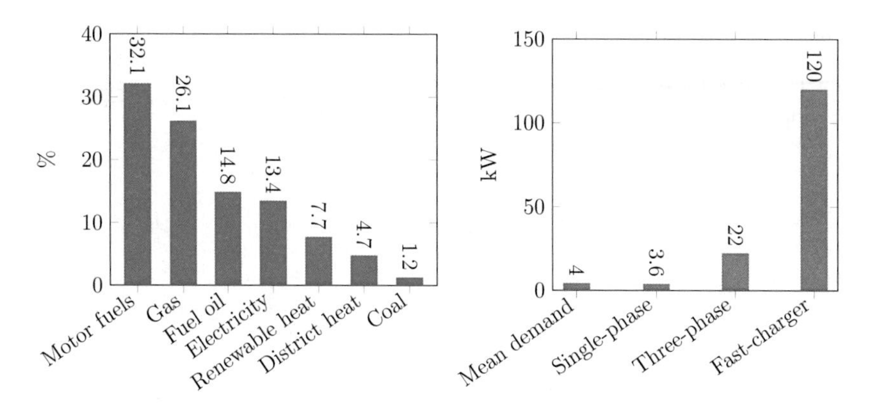

Figure 1.4: Domestic consumption in % of different energy sources (left) and power demand based on a calculated scenario of the BDEW [BDE11] and different power demands by applications (right)

electricity and the demand on a domestic level will grow. Furthermore, the infrastructure needed to provide charging power by installing wall-boxes introduces high power spot loads in the grid [Goe16]. The alternating current (AC) chargers can be designed as single-phase with a rated power of 3.6 kW or connected in all three-phases with a rated power of 11-22 kW [HL17]. For very short charging cycles e.g. at parking places, fast direct current (DC) chargers have been developed, which can have up to several hundred kW [Mou18]. The exemplary power rating of different charger technologies are depicted in Figure 1.4 on the right-hand side.

Electric vehicles can also be used as flexibility, that provides grid to vehicle (G2V) or vehicle to grid (V2G) services. In G2V mode the charging power can be reduced or increased by an external signal, providing a negative power reserve to the operator. In V2G mode the battery of the EV is used to feed active power back into the grid [KT05]. The combination of both modes transforms the EV into mobile battery storage, which can be used to store excess energy from the grid and release it to the grid in high load scenarios.

Negative effects and operational boundaries in low voltage grids

Despite all the advantages that renewable energy sources and electric vehicles offer for our climate, they come with certain technical challenges. Especially, the production of electric energy with smaller renewable DG units like photo-voltaic plants transforms the operation of electrical grids. Their power production strongly depends on weather conditions such as wind and sun. Both short time effects, like wind speed or rapid cloud cover changes, and longtime seasonal differences, result in fluctuations which are only predictable with a certain margin of error [PRK+13]. Furthermore, the production which has its peak during midday and the demand profile in the low voltage levels do not entirely match.

For European distribution systems the standard DIN-EN-50160 regulates the permissible voltage magnitude range and quality requirements for LV grids [DE11]. For three-phase four wire low voltage grids, the line-neutral value of $400\,\mathrm{V}/\sqrt{3}$ and line-line value of 400 V are defined as the nominal value. All three-phase voltages should have a phase shift of 120 ° between two respective phases. The deviation from the nominal voltage magnitude should be smaller than ±10 %, but the range can be extended for long feeders in rural areas up to +10 % and −15 %. These limits should be satisfied for 95 % of the time with an integration interval of 10 min and an observation period of one week.

In addition to the voltage magnitude deviations, four-wire LV distribution grids are subject to unbalances. Even though most households are connected with three-phase four-wire underground cables, the majority of the consumers are still single-phase [REÁC04]. Moreover, single-phase chargers of electric vehicles will add an unbalanced load to this part. On the production side, there are small rooftop photo-voltaic units which are connected to the distribution grids and in-feed active power in one phase only.

According to [DE11], the voltage unbalance caused by these single-phase consumers and by generation in LV distribution grids is restricted to a value of $< 2\%$. On some special occasions, values of 3 % are admissible. The limitation is motivated by the significant impact the voltage unbalance has on the grid infrastructure [AMS13]. If for example, one of the phases is heavily loaded, the magnitude of the voltage of the other two phases increases, which can lead to a violation of certain limits in the particular phase. Another direct effect is the increasing losses in transformers and electrical machinery [FSZ10].

The operational boundaries can be set in direct relation to the effects that PV and electric vehicles have on the power grid as depicted in Figure 1.5. Unobservable over-currents will occur more frequently on the branches of the grid and security measures like fuses are not able to detect them. This effect is due to a combination of high in-feed from PV or other renewable sources and the simultaneous charging of electric vehicles during the afternoon [Lan13]. Especially three-phase chargers with power ratings of 11-22 kW have a huge impact in this scenario. Secondly, the nodal voltage at certain nodes in the grid cannot be kept in the defined range due to high in-feed and low load situations during midday. Another effect is a growing voltage unbalance, due to the introduction of single-phase chargers for private charging points. If several of these chargers are connected to the same phase in one branch of the grid, an increase of the neutral conductor current of the low voltage cables is the outcome.

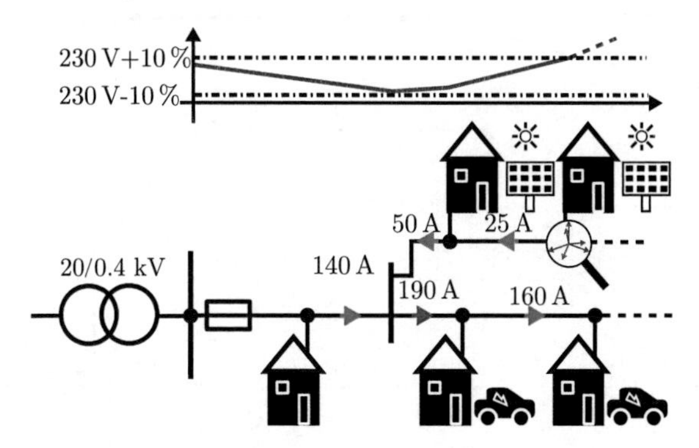

Figure 1.5: Non observable over-current in a line segment of the grid due to the combination of high in-feed and EV charging

1.1.3 From conventional to smart power grids

Smart power grids is a term that summarizes measures to deal with the increasing complexity in the electric power system, especially in the distribution levels. To deal with the increasing complexity, some key technologies can be identified [FMXY12]:

- Communication technology and real-time data acquisition.
- Energy storage devices – in case of LV grids → batteries.
- Intelligent control and operation.

Communication technology and real-time data acquisition

In general the required data set of a power system is collected via measurement equipment in the field and concentrated on a supervisory control and data acquisition (SCADA) system. The specific data is then transmitted to the DSO and informs him of the power system status. In this work it is assumed that either LTE or broadband power line (BPL) are used to transport this data from the measurement equipment in the field to the SCADA system. BPL, which injects a carrier signal modulated on top of the 50 Hz voltage, is regarded to become one of the dominant ways to transmit data in LV grids [AL13].

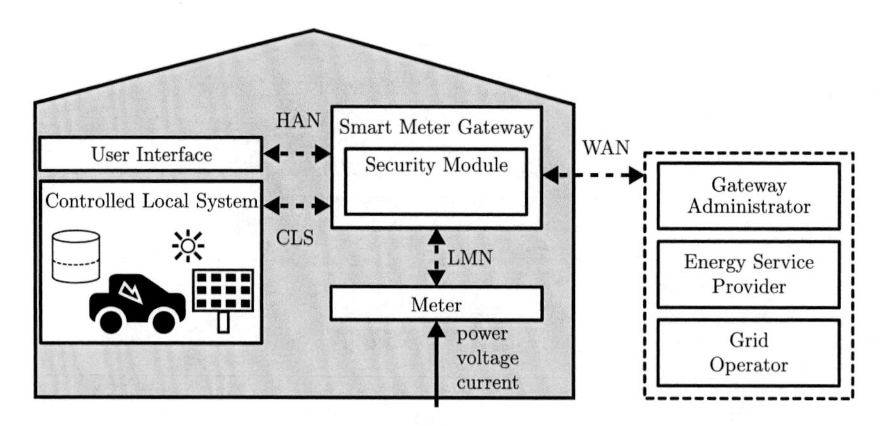

Figure 1.6: Smart metering infrastructure according to the German BSI model

Smart-meters (SM) together with smart meter gateways (SMGWs) will, in future be the dominant way to take care of the acquisition of voltage magnitude, current and apparent power of the monitored nodes [Pep16]. The smart metering infrastructure, as depicted in Figure 1.6 assures the secure acquisition and transportation of collected measurements due to its inherent security features.

Following components form the main part of the smart metering infrastructure:

Local metrological network (LMN) This interface connects the SMGW with measurement devices like electrical or gas meters. Data or information exchanged on this interface is encrypted to assure that measurement data can not be manipulated.

Home area network (HAN) The end user interface enables the consumer to verify the measurements taken by the SMGW for billing. It is only used to view historical measurements for display purposes.

Wide area network (WAN) Through the WAN interface the SMGW is connected to external entities. First of all, they administrate allowed interconnections, set up the pairing to meters in motion and configure end user profiles. Additionally measurements which are used for billing purposes are transmitted through this interface.

Controllable local systems (CLS) Sub-metering devices like heat cost distributors, or controllable devices like PV, battery storages or EVs can be connected to this interface of the SMGW. Through the WAN interface the grid operator, which is an active external market authority (AEMA) can open an encrypted channel to the previously mentioned device at the CLS interface. Through this channel he then can transmit control signals like a curtailment command for PV or collect information.

Gateway Administrator (GWA) Metering point operators administrate the SMGW. The GWA acts as the configuration and maintenance platform for the SMGW. It provides a time signal to synchronize its internal time, starts the interconnection procedure with meters or configures profiles for external entities.

External market authority (EMA) The EMA is an authorized entity which receives measurements regularly collected by the SMGW. In case of an energy service provider the measurements are then processed for billing purposes.

Acive external market authority (AEMA) This type of external entity has additional rights and can control components that are connected to the CLS interface of the SMGW. Consequently, DSOs will be part of this group, as they will need to control PV, EVs or battery storages over this interface.

Energy storages - especially batteries

One key technology for the integration of distributed generation are battery storages for short and long term operations. Firstly, they can be used to overcome the intermittent character of renewable generation units and filter fluctuations short-term. Secondly, they can help to balance energy demand and production long-term. However, implementing large amounts of storage technologies in distribution grids requires a more sophisticated methodology for dispatch and optimization. Energy constraints of the different storage devices have to be considered while putting them to use for different applications. Because of the before mentioned reasons, some works have touched on the subject of increasing the accuracy of functional battery models, the optimal placement, as well as storage scheduling that is necessary for large scale integration [LWX12, For17].

Even though current prices for battery packs are still too high to make them economically interesting for small house owners, the cell prices in Euro/kWh have been decreasing more rapidly than was predicted a few years ago [NN15]. Thus it is safe to assume that soon they will be interesting for domestic consumers and as such a viable source of flexibility in the distribution grids.

Intelligent control and operation

Keeping a high quality of service during transmission and distribution of electric power requires a constant adjustment and active operation of the power grid. Intelligent strategies that utilize the flexibility offered by generation units and loads like EVs, are in the focus of different engineering disciplines [FMXY12, AZG13, GAC+16]. An intelligent control can support the grid with different ancillary services. Four of the main services, which usually have to be provided in a power grid are [DEN17]:

Frequency control Stabilizing the frequency in the power grid is the responsibility of the TSO. In order to achieve this, the balance between load and generation must be maintained at all times. Measures to accomplish this are traditionally the momentary reserve power from rotary masses. If the grid is out of balance for a longer time scale regulating power from another grid zone can be used.

Voltage control In accordance with voltage maintenance regulations, the system operators have to keep the voltage quality in certain predefined regions. An essential factor for voltage maintenance is the provision of reactive power from generation units and network resources.

Operation management Operators have the task of organizing the grid by continuously monitoring and adjusting mechanisms. These adjustments include the dispatch of generation and flexible loads. Furthermore, it is their task to limit constraint violations (e.g., current flow overloads, voltage range) in the power lines or cables.

Grid restoration In the event of a large-scale failure in the power grid, TSOs together with the DSO must be capable of restoring the supply of the electrical grid within a short time. Necessary measures and strategies are bundled in the system service. Not every power plant can take part in grid restoration.

Traffic light concept for distribution grid operation

During the operation of distribution grids, constraints should be kept by all means, to prevent damage to assets and grid equipment. However, considering that different participants like the market, the consumer, or the DSO, have conflicting interests, the right to access components in the smart-grid must be regulated. One proposal for such a regulation was brought together by the German Bundesverband der Energie und Wasserwirtschaft (BDEW) in a discussion paper [Zac17]. The main idea is that there must be a systematic approach and a set of rules as to when the market has full access to the active grid components like PV, batteries, and EV. The solution is that, as long as there is a safe predefined distance maintained with regard to the requirements in [DE11], the market can act free from restriction and use incentives to control distributed generation

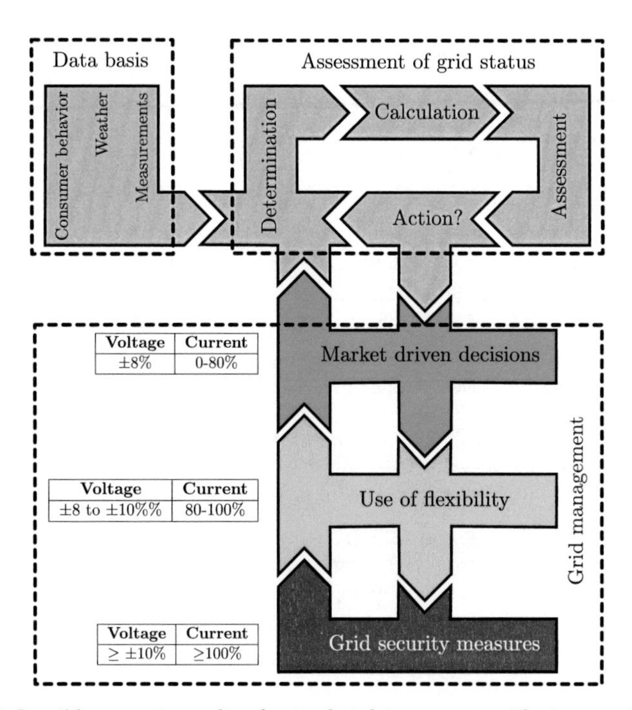

Figure 1.7: Possible operation policy for market-driven smart-grids, in accordance with the traffic light concept by the BDEW [Zac17]

and flexible demand. The different grid states are defined based on a traffic light system as depicted in Figure 1.7. Three main parts in Figure 1.7 describe the procedure of the grid operation from the viewpoint of the DSO:

Data basis In this part, all the measurements available in the grid are concentrated. The data basis consists of measurements collected from a smart-metering infrastructure and the secondary information derived from the measurements. These are load profiles of consumer behavior, prediction of PV generation, and possibly EV charging cycles. These secondary properties can then be used to generate profiles or to substitute missing measurement data.

Grid assessment Based on these predictions and measurements, state estimation algorithms [DWS15, RBT17] are applied to assess the grid state and to identify potential bottlenecks in advance. Whether an action must be taken in grid operation is decided based on this information.

Grid management The grid management is responsible for operation related tasks in the grid. Depending on whether, the assessed grid state is green, yellow or red, different compositions and restrictions of DSO and market take place.

Inside the grid management block, the actual traffic light systems take place in three different stages:

In the green phase the grid is purely market driven. Generation management and the marketing of flexibility is solely based on monetary incentives from external market authorities. Charging of electric vehicles and power management of DG is not restricted by the DSO.

In the yellow phase the grid state is not yet critical but gets closer to it. In this interaction phase the DSO can call upon flexibility, which has been warranted by the market authorities. These authorities receive a compensation in return, for delivering flexible power to the grid. Some of the market decisions from the green phase can be restricted by the DSO if he deems it necessary.

If the red phase is reached and the grid is in a critical state, the DSO immediately applies security measures. This means that active power curtailment of distributed generation or load scheduling is applied. A control signal from DSOs in the red phase always takes priority over the market.

Active operation strategies for low voltage grids

To this day low voltage distribution grids are not operated by the DSO the way medium or high voltages grids are. There is no grid management or operation strategy in place [Oer14, For17, BMTM18]. The reasons for this are manifold, but limited measurement capability and the lack of automation equipment can be seen as significant factors. They pose a high investment cost in relation to the overall length (1.123×10^6 km) and size of the LV grids [DEN18]. During the construction phase LV grids are designed so that consumers and generation units can be safely serviced [Kau95]. During the lifetime of the grid possible bottlenecks are only reevaluated if new customers like photo-voltaic plants or large charging stations for electric vehicles are to be connected [Ker10, Lan13]. Nevertheless, practical installation of demonstration grids show that an implementation is feasible [BAJ13, OLS+13]

Most presented demonstration approaches include decentralized control with only local knowledge [BAJ13]. However, due to the growing number of photo-voltaic plants in the grids this might not be feasible or even lead to unstable conditions [ABK+15]. The introduction of the smart-metering infrastructure will change this. Still, if only local information about states of individual nodes, such as voltage magnitude or power exchange with the grid are available, it is almost impossible for an operation strategy to include the feedback of individual nodes on the entire network. Using load flow techniques which incorporate the network topology and the characteristics of power lines, a much more efficient interaction mapping of individual participants onto the grid can be realized.

A systematic approach that is very promising for distribution grids and at the same time reflects the opinions and goals of the DSOs, was presented with the traffic light concept in Figure 1.7. In the green and yellow phase of the traffic light model, where the market has some freedom to operate the power system, control approaches which make use of locally available flexibility and at the same time satisfy market driven goals have to be deployed. The red phase is only relevant for the security of supply and to prevent grid equipment from damage. For the traffic light concept both central and distributed control mechanisms can be applied. However, only the **operation management** and **voltage control** are in the scope of the traffic light model. These ancillary services are usually sufficient in low voltage distribution systems if no separation from the main medium voltage grid at the substation transformer is required. Thus, in this thesis the **frequency control** or **grid restoration** are not considered further. In the following sections control methods are introduced, that can be integrated for all three traffic light phases.

1.2 Approaches for the optimal operation of power grids

1.2.1 Dynamic optimal power flow

Efficient operational strategies in today's transmission networks usually involve solving an optimal power flow (OPF) [COC12]. The goal of an OPF is to find the best allocation or reduction of certain system relevant variables subject to an objective function and constraints. The objective function can be formulated in various forms such as fuel cost, transmission losses, reactive power injection or unbalance mitigation. It foremost reflects the interests of system operators to reduce operational costs. Solving the OPF includes minimizing an objective function subject to the power flow constraint, the grids transmission capabilities and the limits of generation equipment [FSR12][FR16].

From a grid perspective the solution of the OPF is achieved by adjusting the output power of controllable generation and dispatch-able demand units while satisfying uncontrollable loads. Especially in todays power grids with an increasing amount of storage units intermittent constraints to account for dynamic relationships have to be accounted for. In this case the optimal power flow extends to an dynamic optimal power flow (DOPF) including intermittent constraints and a time horizon [GKA14]. The optimization problem to describe a DOPF can take on the following form

$$\min_{\boldsymbol{p}_k} \sum_{l=0}^{N_{\mathrm{H}}-1} \boldsymbol{p}_{k+l}^{\mathrm{T}} \boldsymbol{C}_{\mathrm{quad}} \boldsymbol{p}_{k+l} + \boldsymbol{p}_{k+l}^{\mathrm{T}} \boldsymbol{C}_{\mathrm{lin}} \boldsymbol{p}_{k+l} + \boldsymbol{C}_{\mathrm{const}} \qquad \text{Fuel costs} \tag{1.1a}$$

s.t.:

$$\boldsymbol{g}\left(\boldsymbol{p}_{k+l}, \boldsymbol{q}_{k+l}, \boldsymbol{u}_{k+l}, \boldsymbol{i}_{k+l}\right) = \boldsymbol{0} \qquad \text{Power flow constraints} \tag{1.1b}$$

$$E_{k+l+1}^{\mathrm{Batt}} = \boldsymbol{A}_{\mathrm{d}} E_{k+l}^{\mathrm{Batt}} + \boldsymbol{B}_{\mathrm{d}} \boldsymbol{p}_{k+l} \qquad \text{Storage dynamics} \tag{1.1c}$$

$$E_{k+l}^{\mathrm{Batt,min}} \leq E_{k+l}^{\mathrm{Batt}} \leq E_{k+l}^{\mathrm{Batt,max}} \qquad \text{Storage limitations} \tag{1.1d}$$

$$\begin{pmatrix} \boldsymbol{p}_{k+l} \\ \boldsymbol{q}_{k+l} \end{pmatrix}^{\mathrm{T}} \boldsymbol{A}^{\boldsymbol{u}} \begin{pmatrix} \boldsymbol{p}_{k+l} \\ \boldsymbol{q}_{k+l} \end{pmatrix} \leq S^{\max} \qquad \text{Generation limits} \tag{1.1e}$$

$$\boldsymbol{u}_{k+l}^{\mathrm{T}} \boldsymbol{A}^{\boldsymbol{u}} \boldsymbol{u}_{k+l} \leq U^{\max} \qquad \text{Nodal voltage limit} \tag{1.1f}$$

$$\boldsymbol{i}_{k+l}^{\mathrm{T}} \boldsymbol{A}^{\boldsymbol{i}} \boldsymbol{i}_{k+l} \leq I^{\max} \qquad \text{Branch current limit} \tag{1.1g}$$

In this form the DOPF is solved for several consecutive time steps $l = \{0, \cdots, N_{\mathrm{H}}\}$ up to some predefined horizon N_{H}. This horizon could be 24 h with discrete time intervals of 1 h. However, a DOPF does not have an integrated feedback from the measurements of the power grid and as such rather acts as a planning tool. Nevertheless, as it uses the nonlinear power flow equations (1.1b) the solution of the full AC DOPF is very accurate.

Advances in the evaluation of the OPF and DOPF has led to efforts solving the problem in a distributed fashion since the late 90s [CA98]. Especially, distributed optimization techniques based on dual decomposition (DD) [CMCB06] and its successors the alternating direction method of multipliers (ADMM) [BPC+11] have been applied to several power system related problems. Even their need for a central coordinator has been solved by an extension called the proximal message passing (PMP) [KCLB13]. Additionally, see [MDS+17] for a complete review on the subject of distributed optimization using different algorithms based on ADMM.

The goal in this thesis is to make use of the accuracy of the full DOPF, while also integrating a feedback from the grid. A control method that uses mathematic optimization to find the best feedback law is presented in the following section.

1.2.2 Model predictive control

System model

It is assumed that the plant under investigation can be described by a linear discrete, time invariant state space equation of the following form

$$\begin{aligned} \boldsymbol{x}_{k+1} &= \boldsymbol{A}_{\mathrm{d}}\boldsymbol{x}_k + \boldsymbol{B}_{\mathrm{d}}\boldsymbol{u}_k \\ \boldsymbol{y}_k &= \boldsymbol{C}_{\mathrm{d}}\boldsymbol{x}_k + \boldsymbol{D}_{\mathrm{ds}}\boldsymbol{d}_k \end{aligned} \tag{1.2}$$

With the discrete system matrix $\boldsymbol{A}_{\mathrm{d}}$ and $\boldsymbol{B}_{\mathrm{d}}$, $\boldsymbol{C}_{\mathrm{d}}$ the input and output matrix. $\boldsymbol{D}_{\mathrm{ds}}$ is the disturbance matrix. The variables $\boldsymbol{x}_k = [x_1, x_2, \cdots, x_N]^{\mathrm{T}}$ $\boldsymbol{u}_k = [u_1, u_2, \cdots, u_N]^{\mathrm{T}}$ and $\boldsymbol{d}_k = [d_1, d_2, \cdots, d_N]^{\mathrm{T}}$ are the discrete state vector, system input vector and measurable disturbance vector respectively. k is the discrete time instant. This state space representation can be found by discretization of a continuous-times states space model, using a zero-order hold (ZOH) with a fixed sample time [Lun16].

Control Approach

In model predictive control (MPC) the actuator signals are calculated by solving an optimization problem. Based on a physical model of the plant, the controller predicts the future behavior subject to possible control actions based on a prediction model (1.3)

$$\begin{bmatrix} \boldsymbol{x}_1 \\ \boldsymbol{x}_2 \\ \vdots \\ \boldsymbol{x}_{N_{\mathrm{p}}} \end{bmatrix} = \underbrace{\begin{bmatrix} \boldsymbol{A}_{\mathrm{d}} \\ \boldsymbol{A}_{\mathrm{d}}^2 \\ \vdots \\ \boldsymbol{A}_{\mathrm{d}}^{N_{\mathrm{p}}} \end{bmatrix}}_{\Phi} \boldsymbol{x}_0 + \underbrace{\begin{bmatrix} \boldsymbol{B}_{\mathrm{d}} & \boldsymbol{0} & \dots & \boldsymbol{0} \\ \boldsymbol{A}_{\mathrm{d}}\boldsymbol{B}_{\mathrm{d}} & \boldsymbol{B}_{\mathrm{d}} & \dots & \boldsymbol{0} \\ \vdots & \vdots & \ddots & \vdots \\ \boldsymbol{A}_{\mathrm{d}}^{N_{\mathrm{p}}-1}\boldsymbol{B}_{\mathrm{d}} & \boldsymbol{A}_{\mathrm{d}}^{N_{\mathrm{p}}-2}\boldsymbol{B}_{\mathrm{d}} & \dots & \boldsymbol{B}_{\mathrm{d}} \end{bmatrix}}_{\Gamma} \begin{bmatrix} \boldsymbol{u}_0 \\ \boldsymbol{u}_1 \\ \vdots \\ \boldsymbol{u}_{N_{\mathrm{p}}-1} \end{bmatrix} \tag{1.3}$$

$$\begin{bmatrix} \boldsymbol{y}_1 \\ \boldsymbol{y}_2 \\ \vdots \\ \boldsymbol{y}_{N_{\mathrm{p}}} \end{bmatrix} = \underbrace{\begin{bmatrix} \boldsymbol{C}_{\mathrm{d}}\boldsymbol{A}_{\mathrm{d}} \\ \boldsymbol{C}_{\mathrm{d}}\boldsymbol{A}_{\mathrm{d}}^2 \\ \vdots \\ \boldsymbol{C}_{\mathrm{d}}\boldsymbol{A}_{\mathrm{d}}^{N_{\mathrm{p}}} \end{bmatrix}}_{\Xi} \boldsymbol{x}_0 + \underbrace{\begin{bmatrix} \boldsymbol{0} & \boldsymbol{0} & \dots & \boldsymbol{0} \\ \boldsymbol{C}_{\mathrm{d}}\boldsymbol{B}_{\mathrm{d}} & \boldsymbol{0} & \dots & \boldsymbol{0} \\ \vdots & \vdots & \ddots & \vdots \\ \boldsymbol{C}_{\mathrm{d}}\boldsymbol{A}_{\mathrm{d}}^{N_{\mathrm{p}}-2}\boldsymbol{B}_{\mathrm{d}} & \boldsymbol{C}_{\mathrm{d}}\boldsymbol{A}_{\mathrm{d}}^{N_{\mathrm{p}}-3}\boldsymbol{B}_{\mathrm{d}} & \dots & \boldsymbol{0} \end{bmatrix}}_{\Psi} \begin{bmatrix} \boldsymbol{u}_0 \\ \boldsymbol{u}_1 \\ \vdots \\ \boldsymbol{u}_{N_{\mathrm{p}}-1} \end{bmatrix} + \underbrace{\begin{bmatrix} \boldsymbol{D}_{\mathrm{ds}} & \boldsymbol{0} & \dots & \boldsymbol{0} \\ \boldsymbol{0} & \boldsymbol{D}_{\mathrm{ds}} & \dots & \boldsymbol{0} \\ \vdots & \vdots & \ddots & \vdots \\ \boldsymbol{0} & \boldsymbol{0} & \dots & \boldsymbol{D}_{\mathrm{ds}} \end{bmatrix}}_{\Theta} \begin{bmatrix} \boldsymbol{d}_0 \\ \boldsymbol{d}_1 \\ \vdots \\ \boldsymbol{d}_{N_{\mathrm{p}}-1} \end{bmatrix} \tag{1.4}$$

With N_p the prediction horizon, which describes how many steps the model is calculated into the future. The controller then chooses the input to the plant $\boldsymbol{u}_k$ based on the minimization of some predefined criteria. The publications [ML99, MRRS00] give a good introduction and overview of the method.

Now to illustrate the overall principle of MPC the main behavior is depicted in Figure 1.8. Because only the basic principle is depicted, a single input single output (SISO) system is shown. It should be noted, that MPC is not limited to SISO systems, but is easily extended for multiple input multiple output (MIMO) systems.

In the given example, a reference trajectory y_k^{ref} should be tracked by the system output y_k. The discrete-time controller, manipulates the control input u_k inside the prediction horizon $l = 0, \cdots, N_{\text{P}} - 1$. The goal of the controller in this example is to minimize the difference to the given reference value $\Delta y_k = y_k - y_k^{\text{ref}}$, which is formulated as its objective function. The system output y_k is measured at time instance k and the controller calculates an optimal future input sequence $u_{k+l}, l = 0, \cdots, N_{\text{P}} - 1$ based on the discrete-time model (1.3) in order to obtain a desired future output trajectory $y_{k+l}, l = 1, \cdots, N_{\text{P}}$ which is closest to the reference trajectory y_k^{ref}[Ber13].

One of the great advantages of MPC is the possibility to consider hard system limita-

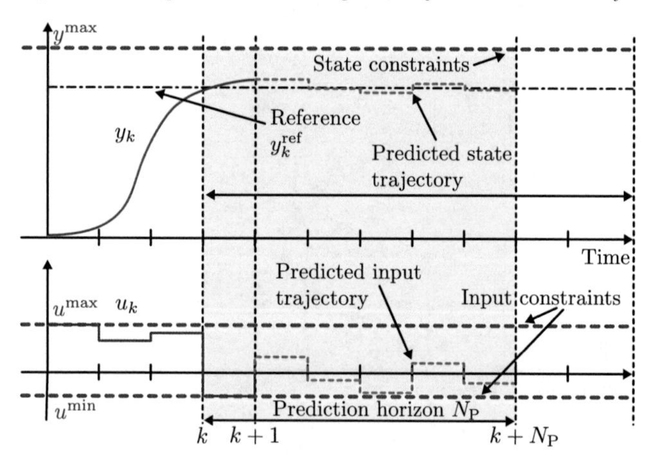

Figure 1.8: Graphical representation of the MPC principal operation

tions on outputs, states or inputs which are further called constraints. The controller predicts the future behavior of the plant and selects the input u_k, such that it stays in defined boundaries. This way, actuator saturation or limitations on the states can be taken into account. This feature enables the MPC to operate closer to system boundaries than in conventional control and consequently leads to a better controller performance [CBA07].

The constraint optimization problem

The core of the MPC is an optimization problem which is formulated by an objective that should be achieved. In the example Figure 1.8 this was the reference tracking. For the remainder of this thesis the optimization problem which has to be solved takes on the form of a quadratically constraint quadratic program (QPQC)[Dom13].

$$\min_{\boldsymbol{u},\boldsymbol{x}} \sum_{l=0}^{N_{\mathrm{P}}-1} \left(\boldsymbol{x}_{k+l}^{\mathrm{T}}\boldsymbol{O}_x\boldsymbol{x}_{k+l} + \boldsymbol{u}_{k+l}^{\mathrm{T}}\boldsymbol{O}_u\boldsymbol{u}_{k+l}\right) + \boldsymbol{x}_{k+N_{\mathrm{P}}}^{\mathrm{T}}\boldsymbol{O}_{\mathrm{P}}\boldsymbol{x}_{k+N_{\mathrm{P}}} \tag{1.5a}$$

s.t.:

$$\boldsymbol{x}_{k+l+1} = \boldsymbol{A}_{\mathrm{d}}\boldsymbol{x}_{k+l} + \boldsymbol{B}_{\mathrm{d}}\boldsymbol{u}_{k+l} \quad \text{System dynamics in } \boldsymbol{x}_k \text{ and } \boldsymbol{u}_k \tag{1.5b}$$

$$\boldsymbol{x}_k = \overline{\boldsymbol{x}}_k \quad \text{Continuity constraint in } \boldsymbol{x}_k \tag{1.5c}$$

$$\boldsymbol{A}_{\mathrm{equ}}^{x}\boldsymbol{u}_{k+l} = \boldsymbol{B}_{\mathrm{equ}}^{x} \quad \text{Equality constraints in } \boldsymbol{x}_k \tag{1.5d}$$

$$\boldsymbol{A}_{\mathrm{equ}}^{u}\boldsymbol{u}_{k+l} = \boldsymbol{B}_{\mathrm{equ}}^{u} \quad \text{Equality constraints in } \boldsymbol{u}_k \tag{1.5e}$$

$$\boldsymbol{A}_{\mathrm{inequ}}^{x}\boldsymbol{x}_{k+l} \leq \boldsymbol{B}_{\mathrm{inequ}}^{x} \quad \text{Inequality constraints in } \boldsymbol{x}_k \tag{1.5f}$$

$$\boldsymbol{A}_{\mathrm{inequ}}^{u}\boldsymbol{u}_{k+l} \leq \boldsymbol{B}_{\mathrm{inequ}}^{u} \quad \text{Inequality constraints in } \boldsymbol{u}_k \tag{1.5g}$$

$$\boldsymbol{x}_{k+l}^{\mathrm{T}}\boldsymbol{A}_{\mathrm{quad},k+l}^{x}\boldsymbol{x}_{k+l} \leq x_{\mathrm{quad}}^{\mathrm{max}} \quad \text{Quadratic constraints in } \boldsymbol{x}_k \tag{1.5h}$$

$$\boldsymbol{u}_{k+l}^{\mathrm{T}}\boldsymbol{A}_{\mathrm{quad},k+l}^{u}\boldsymbol{u}_{k+l} \leq u_{\mathrm{quad}}^{\mathrm{max}} \quad \text{Quadratic constraints in } \boldsymbol{u}_k \tag{1.5i}$$

The constraint optimization problem can be formulated in dense or sparse form or even in a combination, depending on the application and memory requirements of the processing platform [Ken17]. In the dense form, the system states are substituted into the optimization problem based on the prediction model (1.3). This way, the amount of variables in the optimization problem is smaller than in sparse form. Only the dense form will be further investigated, which expresses the system behavior solely based on the system input.

Closed loop operation

The Input sequence calculated based on the optimization problem (1.5a)-(1.5i) is still an open-loop control law. However, if unmeasured or unknown disturbances act on the system, which is not predicted by the model, the input sequence is not optimal anymore.

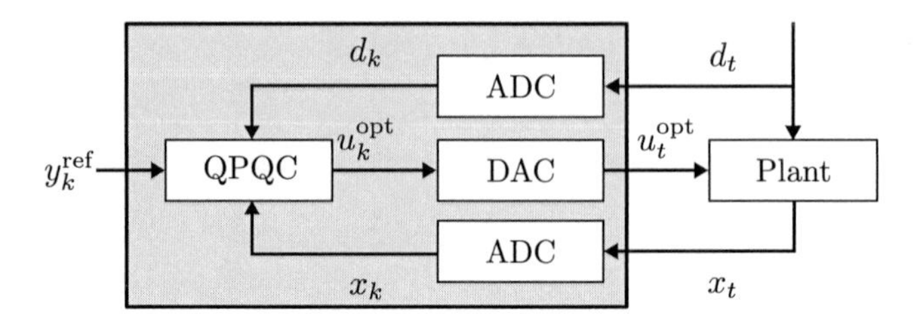

Figure 1.9: MPC closed loop structure (grey box), with the QPQC solver, the analog digital conversion and the plant in continuous time

The reason for this, is that the disturbance was not considered during the computation of the input sequence. To overcome this situation, a feedback of the real system behavior to the controller is introduced. At every time instant k, the states of the system are measured and the optimal input sequence $u_{k+l}^{\text{opt}}, l = 0, \cdots, N_{\text{P}} - 1$ is calculated with the knowledge of the real system state. Instead of using the whole optimal input sequence, only it's first element u_k^{opt} is applied to the system.
The system states are measured at every time instant k and the prediction-horizon moves forward after each computation cycle. Controllers can receive new informations about the system, disturbances or changing parameters. Due to this behavior MPC is also called Receding Horizon Control (RHC).
As already mentioned currently the controller works in discrete time while the plant is a continuous time process. Thus the control is interfaced to the system via analog digital and digital analog converters (ADC), (DAC) (see Figure 1.9). With u_k^{opt} the optimal control signal as an input to the plant. Variables with index t are in continuous time.

Up to now only a single MPC that controls the overall plant was discussed. For large systems the solution of the optimization problem in this centralized setup may consume a lot of processing power. Furthermore, interaction between the MPC and sensors and actuators can be quite challenging. So in order to make this approach more suitable for larger applications, MPC has been further extended to a distributed version called distributed MPC (DMPC).

1.2.3 Distributed model predictive control

Because of the size and complexity of power systems, a single central controller is not regarded as feasible for operational purposes with MPC [HJL+12]. Thus the concept of distributed control is introduced in the following part. The general idea behind dis-

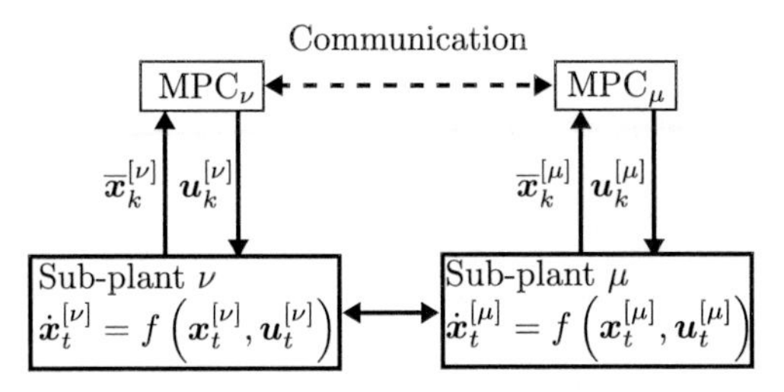

Figure 1.10: Distributed MPC setup with to local controller interacting with local subsystems

tributed model predictive control (DMPC) is quite simple – the central large scale control problem, with all it's sensory inputs and actuation possibilities, is divided into several parts (subsystems). Figure 1.10 shows a principal setup with two local controllers and sub-plants. The non-overlapping but interconnected subsystem are then each assigned to a local MPC controller [JK01]. DMPC features many advantages, such as resilience to controller faults, because the overall system is operated with many local controllers that can act independently. Furthermore, the grid controller maintenance can be simplified due to possible modularity and scalability. But especially the resilience to faults makes distributed schemes interesting for power systems, with very high security of supply demands.

Physical coupling of the local subsystems and the coordination method of the controllers have a strong interdependency [Moh13]. Additionally, there are several possibilities as far as how the controllers can interact. They can share a common objective which they want to minimize or use coordination for stability reasons [CJKT02]. If necessary, it is possible to define an objective function that emulates a centralized controller, as was presented in [Ven06]. Consequently, the type of operation and cooperation defines the goal of each DMPC algorithm. From the literature, two principal directions can be identified. On the one hand, some algorithms solve the central problem based on distributed optimization. A process of several iterations between each of the subproblems is needed until convergence is reached [MN14]. On the other, some methods do not make use of distributed optimization, but rather use robust control techniques [MSR05, Sca09, CSdlPL13].

One popular example of a robust scheme is tube-based MPC [CZMJ13]. The main idea is the introduction of an off-line calculated feedback law that binds the states of each subsystem inside an invariant set. This feedback term is added to the optimized system input. An upper bound can always calculate the worst-case interaction between subsystems. Through this feedback, the effect of the individual disturbances is minimized. Robust distributed MPC is not the subject of this thesis and from now on only iterative DMPC will be further discussed.

Two very prominent examples are communication-based DMPC [VHRW08] and the so-called feasible cooperation based DMPC [VRW06]. In both of the approaches, the local optimization problem is solved individually in a decentralized fashion. After each solution the current optimal control input $u_k^{[\nu]}$ is exchanged. This procedure is repeated within each sampling time of the DMPC to converge to a global equilibrium point. It should be noted that local problems are formulated in a specialized way, that allows for each solution of the DMPC to be feasible. Every additional iteration, for which exchange of information between every local controller is performed, draws the solution further to optimality. In return, this means that after every iteration, each local controller possesses a solution that can be applied to the plant without any risk of constraint violation.

Usually, more recent iterative techniques include dual decomposition, gradient descent, or alternating direction method of multipliers (ADMM) techniques to solve the DMPC problem [GDK$^+$13, Gie13, FSJ14, Mon18]. This family of distributed optimization techniques is not used in this thesis due to their high communication demand.
In the formulation of the DMPC, it is assumed that every subsystem can be coupled with another one via state x, or input u. The local distributed MPC problem (DMPC) ν now has the following form

$$\min_{\boldsymbol{u},\boldsymbol{x}} \quad \sum_{l=0}^{N_\mathrm{P}-1} \left(\boldsymbol{x}_{k+l}^{[\nu],\mathrm{T}} \boldsymbol{O}_x^{[\nu]} \boldsymbol{x}_{k+l}^{[\nu]} + \boldsymbol{u}_{k+l}^{[\nu],\mathrm{T}} \boldsymbol{O}_u^{[\nu]} \boldsymbol{u}_{k+l}^{[\nu]} \right) + \boldsymbol{x}_{k+N_\mathrm{P}}^{[\nu],\mathrm{T}} \boldsymbol{O}_\mathrm{P}^{[\nu]} \boldsymbol{x}_{k+N_\mathrm{P}}^{[\nu]} \tag{1.6a}$$

$$\text{s.t.:} \quad (1.5\text{d}) - (1.5\text{i}) \quad \text{For each subsystem } \nu$$

$$\boldsymbol{x}_{k+l+1}^{[\nu]} = \boldsymbol{A}_\mathrm{d}^{[\nu]} \boldsymbol{x}_{k+l}^{[\nu]} + \boldsymbol{B}_\mathrm{d}^{[\nu]} \boldsymbol{u}_{k+l}^{[\nu]} + \underbrace{\sum_{\mu=1,\mu\neq\nu}^{N-1} \boldsymbol{A}_\mathrm{d}^{[\mu]} \boldsymbol{x}_{k+l}^{[\mu]} + \boldsymbol{B}_\mathrm{d}^{[\mu]} \boldsymbol{u}_{k+l}^{[\mu]}}_{\text{Coupling with subsystem } \mu} \tag{1.6b}$$

$$\boldsymbol{x}_k^{[\nu]} = \overline{\boldsymbol{x}}_k^{[\nu]} \tag{1.6c}$$

The additional term in the system dynamic (1.6b) accounts for the interaction of the subsystems. In this example the objective is only a function of the local subsystem states and inputs. As discussed before the cost function can also include cost terms from other subsystems [Ven06].

Advantages and disadvantages of model predictive control

Model predictive control is a high performance and advanced control method. The possibility to adjust this performance based on an objective function gives the control designer a certain amount of freedom. As the method uses a system model to predict the plant behavior, measurable and observable disturbances can be integrated in the prediction. Furthermore, system limitations in the form of constraints can be easily integrated and considered by the controller. MPC is used for broad range of different system classes (time-varying, non-linear, stochastic in both MIMO and SISO). Concepts to prove stability exist and are usually based on a combination of invariant sets and a final weighting term [MRRS00]. However, the complexity of the algorithms to solve the constraint optimization problem inside the MPC is computationally expensive [Ken17]. Furthermore, the prediction of the plant behavior requires an accurate model of the underlying system. This means that modeling techniques and system parameter identification play a vital role in acquiring the model.

Model predictive control is a good framework for integrating dynamic optimal power flow. It already uses the same structure of a constraint optimization problem, but adds feedback from the power system to mitigate the effect of disturbances and parameter uncertainties. Especially distributed model predictive control and its properties are very well suited for grid operation purposes. The main advantages are

- Resilience to controller failure.
- Adaptation to grid changes $\rightarrow$ plug and play.
- Smaller subproblems, less computationally demanding.

Even though the distributed MPC methods presented in literature can handle a wide range of convex constraints, integrate data security aspects and have good convergence properties, the main disadvantage however is the huge amount of communication that is necessary for the optimization problem to converge.

1.3 Contribution of this work

The secure on-line deployment of an operational strategy for power grids based on distributed control, requires methods that need limited communication and can assure feasible solutions even after a single iteration. This way communication failures are not severe for the grid security and only lead to suboptimal solutions. Instead of relying on a strong communication infrastructure, as is done for example in ADMM, an approach is chosen that assumes good knowledge about the underlying power system. Hence, the necessary communication and interaction between local controllers is kept to a minimum. The approach developed in his work can thus be classified as a feasible cooperation, for which the main concept has been introduced in [Ven06]. However, in an addition to previous work, state constraints can be integrated in the optimization problem. As a consequence each developed operation strategy needs a model of the power grid.

Using a system-wide grid model for each controller increases the local computational effort, which is not desired. This disadvantage is efficiently handled with the developed power flow approximation, called the linear interpolation power flow (LIPF). Instead of using an approximation method e.g. Taylor [CT08], a linear function with variable parametrization is derived from the Forward-Backward-Sweep power flow solution method [Zim95]. The parameters of the function are calculated based on acquired measurements of local apparent power and voltage. The repeating feedback of measurements into the distributed optimization and the on-line calculation of parameters for the LIPF are synchronized. Hence, for each prediction cycle of the controllers a simple and accurate representation of the power grid is available. Additionally, the phase angles in LV grids are small [For17] and thus voltage magnitude measurements are sufficient to parametrize the LIPF, which makes the use of phase measurement units (PMUs) unnecessary. The smart-meter infrastructure depicted in Figure 1.6 is utilized to receive local measurements of power and voltage magnitude from base meters.

For the overall operation management several layers of distributed algorithms are implemented, which allow for different necessary features like reactive power support, congestion management, unbalance mitigation and parameter estimation to be integrated in one common structure. By using the developed approximated power flow for all three approaches a high level of re-usability for the algorithms can be achieved as depicted in Figure 1.11.

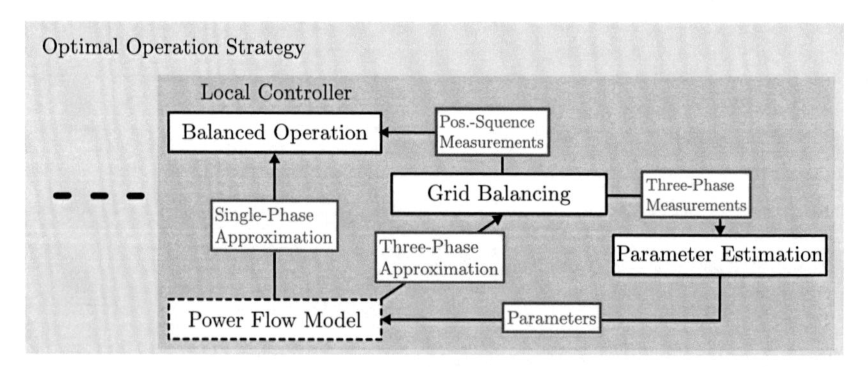

Figure 1.11: Different contributions of this thesis, which interlink with each other to form the optimal operation of the grid

The first strategy assumes that the grid can be modeled as a balanced three-phase system. It implements voltage control and operation management in a distributed MPC scheme. Usually, the controllers would exchange the power and voltage measurements directly to find a common solution. Even though information channels between the controllers can be encrypted, and thus attacks from the outside can be minimized, each controller would still know the power trajectories of the others. To overcome this challenge and generate a certain amount of privacy, every local controller constructs a power chart for each time instant in the prediction horizon of the control. These power charts reflect the flexibility each node can offer for the grid operation, without compromising their sensitive data, e.g., battery storage or charger size. The local controllers then share these power charts over a common communication bus to jointly decide for a dispatch solution. This minimizes grid losses and at the same time respects the local interest of each controller. With the help of simulation studies, the advantage of this method is evaluated with different benchmark grids. Furthermore, a demonstrator based on a hardware power grid emulation and a rapid prototyping system was developed to show that the algorithm can be implemented on a real target computer.

The second strategy tackles the inherent unbalance in low voltage grids that can induce damage in different assets. The mitigation algorithm is formulated as a distributed, dynamic optimal power flow problem. By utilizing the linear and explicit character of the LIPF, it can be exploited to decompose the three-phase optimal power flow into three linear independent system descriptions. This is a stark contrast to the standard transformation of the full AC optimal power flow into symmetrical components [Das17, BLR+11, APCP13]. Due to the simplicity, a Jacobi type algorithm is sufficient to solve the three-phase OPF in a distributed fashion cooperatively. Simulation studies with a 40 low voltage node benchmark grid show that controllable single-phase loads have the largest influence in mitigating the unbalance. Furthermore, the analysis of the results show that in low voltage grids the zero-sequence component has a much larger influence on the unbalance than the negative-sequence.

Power grids can change due to installations of new underground cables or exchange of substation transformers. As a consequence, the underground cables in the power flow model of the grid are different from before. Previous works have approached the problem and solved the static mean value estimation of grid parameters [LK09, PGRB16b, PGK+16]. However, the impedance of the cables do not only change due to constructional extensions of the grid but have a thermal dependency related to the branch-current flow as well. This interdependency increases the resistance of the cables significantly and influences the power flow model used for the grid assessment. A thermal dependency model, together with the cable parameters, was formerly utilized to rate the capacity of the grid [Mil06, Deg15]. In this work, the two approaches are combined, and an automatic mechanism to integrate the parameter changes into the operational strategies is developed.
Furthermore, it has to be considered that there is a strong unbalance in the nodal volt-

age and branch currents during the collection of measurement data from smart-meters. A novel masking process uses an indicator based decision approach to filter which measurements are suitable for the parameter estimation. The changing parameters act as feedback into the power flow approximation, which is used inside the two operational strategies. Studies with an unbalanced benchmark grid show that this approach increases the accuracy of the voltage and current estimation significantly.

1.4 Thesis overview

Chapter 2: first introduces the underlying mathematical model of the power system and gives a short review of the state of the art power flow algorithms. Then the general relationship between apparent power and nodal voltage is further discussed, and the link to voltage stability analysis is introduced. After this, a literature review and comparison of power flow approximations is introduced, that is followed by a brief description of the approximated power flow (LIPF) developed in this work. An in-depth investigation of the properties of the LIPF is provided. This is achieved by comparing it with state of the art methods from literature in a simulation.

Chapter 3: introduces the benchmark grids which are used for the evaluation of the different strategies. This includes the well known CIGRÉ low voltage benchmark grid and one that was derived from a real rural low voltage power grid. Furthermore, a smart-grid demonstrator which was developed for the verification of the balanced operation algorithm is introduced and described.

Chapter 4: the distributed control strategy for the balanced operation of the LV grid is introduced, and its main features are validated using the benchmark grids in a simulation. This simulation provides insight into storage management, and constraint satisfaction. Furthermore, the practical feasibility of the approach is provided by a real-time implementation on a dSpace rapid-prototyping system coupled with an emulated 40 node low voltage grid. Through a virtual substation, the emulated grid is connected to a scaled medium voltage power grid demonstrator.

Chapter 5: develops a three-phase balancing controller which is based on a distributed optimization with a Jacobi type algorithm. The controllers are coupled to the operational controller from Chapter 4 and evaluate the full three-phase power flow to estimate the unbalance in the grid. Based on the LIPF developed in Chapter 2, an explicit formulation of the voltage sequence-components is achieved, allowing for them to be minimized in a separated or unified manner. The distributed algorithm is investigated with a simulation using a highly unbalanced power grid, showing how different loads and generations units in the grid can be deployed for an efficient balancing.

Chapter 6: introduces the parameter estimation algorithm, which is specifically designed for highly unbalanced power grids. It introduces the procedure which combines the estimation of impedance mean values with an on-line dynamic thermal model. This influence of changing resistance parameters of the power cables due to thermal heating is then quantified for the LIPF with a highly unbalanced scenario from Chapter 5.

Chapter 7: Summarizes the thesis and briefly discusses the results. Moreover, an outlook to further developments and possible future directions are given.

2 Power system modeling and efficient power flow approximations

The efficient solution of the power flow equations has ever since been a field of active research. This chapter first gives a short introduction to the non-linear power flow solution, constructs the underlying power grid model used throughout the thesis and summarizes the efficient state-of-the-art power flow approximations. The last part introduces an explicit power flow approximation which was developed in this work and was partly presented in [BBL17, BBL18]. This explicit power flow approximation called the linear interpolation power flow (LIPF) is compared with its main competitors found in literature. The chapter closes with an evaluation of the accuracy of the proposed LIPF.

2.1 Introduction

In the planning phase of a power system, a power flow analysis is used upfront to determine the steady-state currents and voltages to which the equipment will be exposed to. During the operation of the grid, it can be used as an instrument, to test if the installed equipment, e.g. cables, do not take any damage by a current load situation. Furthermore, it is used to recalculate planned changes to the grid. The calculation of the power flow is performed in a steady state with the use of algebraic equations, as this reflects the dominant behavior of a power system [Mil10].

One of the first methods used to solve the AC power flow was the Gauss-Seidel (GS) method. At the time it had a relatively slow convergence rate but could be calculated with the available hardware [Woo96]. With advances in the development and manufacturing of integrated circuits, more powerful computational units could be built. Especially large matrix inversion could now be handled in acceptable time, which lead to a domination of the Newton-Raphson (NR) method. It's superior convergence properties and ability to handle a wide range of problems, made it the "Swiss Army knife" for most power flow calculations. Additionally, see [Sto74] for a comprehensive review on the historical development of power flow calculations.

2.1.1 Structural properties of power grids and the effect on power flow solutions

Our power system consists of different voltage levels, each having a distinct purpose for the transmission and distribution of electrical power. High voltage grids are used to transmit power over long distances, while the distribution of electricity to the consumers is usually the responsibility of medium and low voltage grids. There are structural as well as parameter differences between the levels. High voltage transmission grids usually have a meshed character, which means that there can be multiple paths for an electrical current to flow between two buses. Other common grid structures are rings and trees, which are mostly found in distribution grids [Ker12]. Especially low voltage grids in suburban and rural ares have either a radial or an open ring structure. The three discussed grid structures are depicted in Figure 2.1.

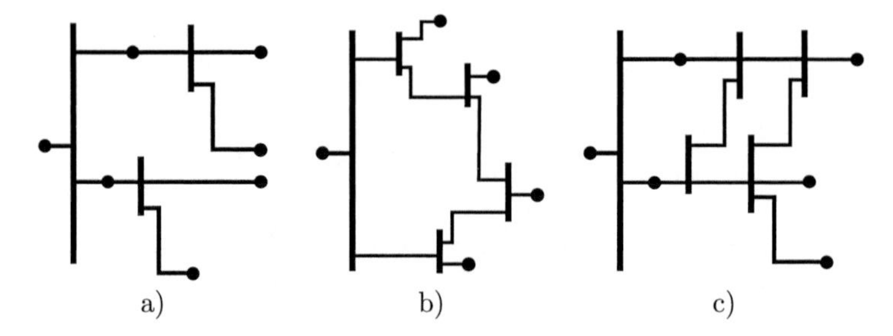

Figure 2.1: Example grids with a) radial, b) ring and c) meshed structure

The difference in structure can be effectively exploited for the power flow calculation. In this context, a drawback of the NR method emerges, as it struggles to handle small X/R ratios, that are common in distribution levels. As a result, the Jacobian matrix which forms a core element in the method is badly conditioned. This, in turn, causes numerical problems if the matrix is inverted. Consequently, it was discovered very early that the conventional NR-method suffers from convergence problems in distribution power flow analysis [SHSL88]. To overcome this problem, specialized methods which exploit the radial structure of distribution grids and are not affected by small X/R ratios have been developed in the past decades. The most notable and widespread of these formulations is the Forward-Backward-Sweep (FBS) method [Zim95]. This method uses the impedance matrix instead of the admittance matrix and can be calculated without matrix inversion. In order to formulate the power flow equations, a mathematical model is needed which relates currents, voltages, and power to each other. In the next section, this mathematical model is introduced with a focused formulation on radial grids.

2.2 Modeling the power grid

2.2.1 Simplified balanced single-phase network model

Distribution systems consist of three phases and a neutral wire or earth conductor. If only consumers with equal loading in each phase are connected to the grid, it is often referred to as balanced conditions. As all three phases have the same loading in active and reactive power their relative behavior with regard to voltage and currents in magnitude and phase is also equal. Thus, the three-phase system can be displayed via an equivalent single phase circuit.

Low voltage cables account for about 80 % of the connections in grids today [Lan13]. Their electrical behavior is dominated by the longitudinal copper resistance and inductance of the conductor. The capacitance induced by the polyvinyl chloride insulation material can be approximated with $0.4\,\frac{\mu\mathrm{F}}{\mathrm{km}}$ [Yin11]. In a Π equivalent circuit this would result in a shunt impedance of $\approx 16\,\mathrm{k}\Omega$ for a line of 1 km length. At the same time, the longitudinal impedance is in a range of mΩ. This shows that any mutual impedance due to the line-line or line-PEN capacitance of the cable can be neglected [Oer14]. So further in this section, only the longitudinal impedance of the cable is considered. Starting from a simplified cable model in a balanced single-phase equivalent representation [Ker12] following setup, depicted in Figure 2.2, is found.

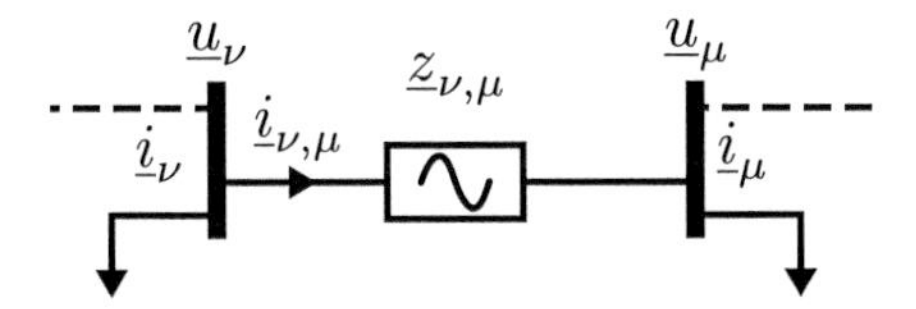

Figure 2.2: Example cable branch notation for a balanced single-phase equivalent circuit

With the branch impedance, $\underline{z}_{\nu,\mu} = r_{\nu,\mu} + jx_{\nu,\mu}$ a combination of the series resistance and inductance of the cable and j the imaginary number. $\underline{u}_\nu, \underline{u}_\mu$ are the complex nodal voltages and $\underline{i}_\nu, \underline{i}_\mu$ are the complex nodal currents. $\underline{i}_{\nu,\mu}$ denotes the branch current between node ν and μ. The indexing of the nodes is done consecutive from the root node (substation transformer) down to the leaves, which represent the last node of a branch.

Inclusion of topology information

The power flow is based on a physical model of the underlying power system described by Kirchhoff's laws. In the case of radial grids investigated here, their topology can be described by graph theory as a tree defined by $\mathcal{G}(\mathcal{N}, \mathcal{O})$. $\mathcal{N}$ is the set of nodes and $\mathcal{O}$ is the set of edges in the graph. An edge stands for a physical connection between two respective nodes and is weighted with an impedance value. The nodes are indexed as $\nu \in \mathcal{N} = \{1, \ldots, N\}$, where N is the number of nodes in the grid.
To combine the branch impedance of the cables or power lines with the topology of the grid the respective inverse branch impedances (Figure 2.2) are stacked in ascending numbering to obtain a diagonal admittance matrix. If this matrix is now right and left multiplied with the incidence matrix $\boldsymbol{W}$ of the network the branch admittance matrix of the grid can be found [Tit11, Woo96]

$$\underline{\boldsymbol{Y}} = \boldsymbol{W}^{\mathrm{T}} \mathrm{diag}\left(\frac{1}{\underline{z}_{1,2}}, \cdots, \frac{1}{\underline{z}_{N-1,N}}\right) \boldsymbol{W}. \tag{2.1}$$

A unique impedance matrix $\underline{\boldsymbol{Z}}$ can be derived from this admittance matrix $\underline{\boldsymbol{Y}}$. This is due to the fact that radial networks have no meshes and thus the nodal currents can be uniquely matched to the respective branch currents. The impedance matrix $\underline{\boldsymbol{Z}}$ can be constructed by cutting the first column and row of the Admittance matrix (2.1) and inverting the resulting sub-matrix. For the mathematical proof of this approach see [BCCC11].
If now the voltage $\underline{u}_0$ is assumed at the root node of the grid (substation transformer) following representation between nodal currents and voltages can be formulated.

$$\underbrace{\begin{pmatrix} \underline{u}_1 \\ \vdots \\ \underline{u}_N \end{pmatrix}}_{\underline{\boldsymbol{u}}} = \underline{\boldsymbol{Z}} \underbrace{\begin{pmatrix} \underline{i}_1 \\ \vdots \\ \underline{i}_N \end{pmatrix}}_{\underline{\boldsymbol{i}}} + \mathbf{1}\underline{u}_0 \rightarrow \underline{i}_0 = \sum_{\nu=1}^{N} \underline{i}_\nu \tag{2.2}$$

With $\mathbf{1}$ being a row vector with N elements equal to 1, corresponding to the number of nodes in the grid. The current $\underline{i}_0$ through the root node can be calculated as the sum of the nodal currents (2.2).
Even though the impedance matrix no longer has a sparse structure, which is the case for $\underline{\boldsymbol{Y}}$ in radial grids [BD15], the explicit formulation with the impedance matrix $\underline{\boldsymbol{Z}}$ has a lot of beneficial properties.

2.2.2 Four-wire network model

For the balanced operation of the power grid, a single phase equivalent representation is sufficient to model the network. However, European low voltage distribution grids

usually are equipped with four-wire cables and overhead lines both in a TN-C-S configuration until after the domestic junction boxes in the houses [PHS05]. The lines have three phases and a PEN conductor with an impedance $\underline{z}^{\mathrm{n}}_{\nu,\mu}$ equal to the line.
The effect of an additional local grounding impedance $\underline{z}^{\mathrm{g}}_{\nu,\mu}$ on the neutral conductor current is usually not that significant, because the resistance of earth can be approximated with 100 Ωm [Ker08, Ben15] which is much higher than the PEN wire from the cable. However, it still reduces the voltage offset between the ground and the neutral conductor which is favorable.

Residential loads and some smaller photo-voltaic units are connected to a single phase of the network. To account for this the single-phase circuit of Figure 2.2 needs to be extended to a three-phase four-wire representation in Figure 2.3. Additionally, shunt impedances are included in the formulation to account for the behavior of local loads. It

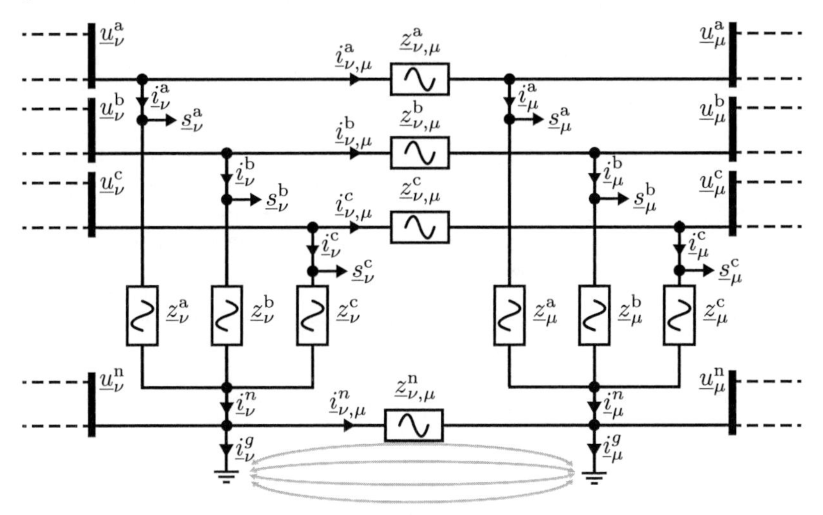

Figure 2.3: Respective equivalent branch circuit for a four-wire low voltage cable, with shunt impedance in star connection accounting for loads

is furthermore assumed that the investigated low voltage cables have symmetrical longitudinal impedances and that the mutual magnetic line to line coupling is neglect-able [Lan13]. Based on this assumption the branch impedance matrix for the three-phase four wire case $\underline{\mathbb{Z}}_{\mathrm{abc}}$ and the PEN conductor impedance matrix $\underline{\mathbb{Z}}_{\mathrm{n}}$ can be constructed separately based on the same procedure as (2.1). This leads to the following line impedance matrix

$$\underline{\mathbb{Z}}_{\text{abc}} = \begin{pmatrix} \boldsymbol{I}^{3\times3}\underline{z}^{\Sigma}_{1,1} & \cdots & \boldsymbol{I}^{3\times3}\underline{z}^{\Sigma}_{1,N} \\ \vdots & \ddots & \vdots \\ \boldsymbol{I}^{3\times3}\underline{z}^{\Sigma}_{N,1} & \cdots & \boldsymbol{I}^{3\times3}\underline{z}^{\Sigma}_{N,N} \end{pmatrix} \tag{2.3}$$

with the value $\underline{z}^{\Sigma}_{1,1}$ equivalent to the entry (1,1) of the balanced impedance matrix $\underline{\boldsymbol{Z}}$, which is a summation out of several branch impedances. These values are then mapped on every phase of the respective node with a 3×3 unity matrix $\boldsymbol{I}^{3\times3}$.

Based on the extended circuit representation of Figure 2.3 the neutral conductor of wye connected loads and shunt impedances is introduced by the following current relation

$$\underline{i}^{\text{n}}_{\nu} = \underline{i}^{\text{a}}_{\nu} + \underline{i}^{\text{b}}_{\nu} + \underline{i}^{\text{c}}_{\nu}. \tag{2.4}$$

Notably, Kron's reduction [CFO03] is not necessary because the mutual line-line and line-PEN impedances are neglected. This reduces the impedance matrix to 3×3. The neutral conductor voltage and current can be fully recovered. However, the Kron reduction method would also not be useful here, as in four-wire grids the voltage drop across the PEN conductor cannot be neglected.

With (2.4) the PEN conductor impedance matrix resulting in

$$\underline{\mathbb{Z}}_{\text{n}} = \begin{pmatrix} \boldsymbol{L}^{3\times3}\underline{z}^{\Sigma}_{1,1} & \cdots & \boldsymbol{L}^{3\times3}\underline{z}^{\Sigma}_{1,N} \\ \vdots & \ddots & \vdots \\ \boldsymbol{L}^{3\times3}\underline{z}^{\Sigma}_{N,1} & \cdots & \boldsymbol{L}^{3\times3}\underline{z}^{\Sigma}_{N,N} \end{pmatrix} \tag{2.5}$$

where $\boldsymbol{L}^{3\times3}$ is a 3×3 matrix with all entries equal to 1. This simplified formulation results from the assumed symmetry properties of low voltage cables impedances. Furthermore, in (2.3) the matrix $\boldsymbol{I}^{3\times3}\underline{z}^{\Sigma}_{N,\mu}$ represents the influence of the currents of node $\mu \rightarrow \underline{i}^{\text{a}}_{\mu}, \underline{i}^{\text{b}}_{\mu}, \underline{i}^{\text{c}}_{\mu}$ on the voltages of the last node N in all phases. Here $\boldsymbol{L}^{3\times3}\underline{z}^{\Sigma}_{N,\mu}$ represents the effect of a PEN conductor current of node μ on the respective voltage drops for each phase of node N.

(2.5) is now integrated with the extended three-phase four-wire representation of (2.2) by a simple summation of the matrices $\underline{\mathbb{Z}}_{\mathrm{abc}}$ and $\underline{\mathbb{Z}}_{\mathrm{n}}$

$$\begin{aligned}\underline{\boldsymbol{\mathcal{U}}} &= \underbrace{\left(\underline{\mathbb{Z}}_{\mathrm{abc}} + \underline{\mathbb{Z}}_{\mathrm{n}}\right)}_{\underline{\mathbb{Z}}_{\mathrm{abcn}}} \underline{\boldsymbol{\mathcal{I}}} + \mathbf{1} \otimes \underbrace{\begin{pmatrix} 1 \\ \underline{m} \\ \underline{m}^2 \end{pmatrix} \underline{u}_0}_{\underline{\boldsymbol{\mathcal{U}}}_0} \\ \underline{\boldsymbol{\mathcal{I}}}_0 &= \sum_{\nu=1}^{N} \underline{\boldsymbol{\mathcal{I}}}_{\nu},\ \underline{i}_0^{\mathrm{n}} = \sum_{\nu=1}^{N} \underline{i}_{\nu}^{\mathrm{n}} \\ \underline{m} &= -\frac{1}{2} - j\frac{\sqrt{3}}{2},\ \underline{m}^2 = -\frac{1}{2} + j\frac{\sqrt{3}}{2}\end{aligned} \tag{2.6}$$

with $\underline{\boldsymbol{\mathcal{I}}} = \left[\underline{\boldsymbol{\mathcal{I}}}_1^{\mathrm{T}}, \cdots, \underline{\boldsymbol{\mathcal{I}}}_N^{\mathrm{T}}\right]^{\mathrm{T}}$ and $\underline{\boldsymbol{\mathcal{U}}} = \left[\underline{\boldsymbol{\mathcal{U}}}_1^{\mathrm{T}}, \cdots, \underline{\boldsymbol{\mathcal{U}}}_N^{\mathrm{T}}\right]^{\mathrm{T}}$ the stacked complex current and voltage vectors of each node ν with $\underline{\boldsymbol{\mathcal{I}}}_{\nu} = \left[\underline{i}_{\nu}^{\mathrm{a}}, \underline{i}_{\nu}^{\mathrm{b}}, \underline{i}_{\nu}^{\mathrm{c}}\right]^{\mathrm{T}}$ and $\underline{\boldsymbol{\mathcal{U}}}_{\nu} = \left[\underline{u}_{\nu}^{\mathrm{a}}, \underline{u}_{\nu}^{\mathrm{b}}, \underline{u}_{\nu}^{\mathrm{c}}\right]^{\mathrm{T}}$. $\otimes$ denotes the Kronecker product [BV04].

The shunt impedances that have been neglected for the construction of the impedance matrix can be summarized as $\underline{\mathbb{Z}}_{\mathrm{sh}} = \mathrm{diag}\left(\underline{z}_1^{\mathrm{a}}, \underline{z}_1^{\mathrm{b}}, \underline{z}_1^{\mathrm{c}}, \ldots \underline{z}_N^{\mathrm{a}}, \underline{z}_N^{\mathrm{b}}, \underline{z}_N^{\mathrm{c}}\right)$ for every respective node. The no load solution with this shunt impedances can be calculated as

$$\underline{\boldsymbol{\mathcal{U}}}_{\mathrm{L0}} = \underline{\mathbb{Z}}_{\mathrm{abcn}} \left(\boldsymbol{I}^{3N \times 3N} - \underline{\mathbb{Z}}_{\mathrm{abcn}} \underline{\mathbb{Z}}_{\mathrm{sh}}^{-1}\right)^{-1} \underline{\mathbb{Z}}_{\mathrm{sh}}^{-1} \mathbf{1} \otimes \underline{\boldsymbol{\mathcal{U}}}_0 + \mathbf{1} \otimes \underline{\boldsymbol{\mathcal{U}}}_0. \tag{2.7}$$

Load models for demand and distributed generation

With the physical model of the power network based on Kirchhoff's current law, a relationship between voltages and currents is defined. In order to describe the power exchange between the nodes, apparent power is introduced as the product of voltage and the complex conjugate of current.

$$\text{Balanced equivalent} \rightarrow \underline{\boldsymbol{s}} = \mathrm{diag}\left(\underline{\boldsymbol{u}}\right) \underline{\boldsymbol{i}}^* = \boldsymbol{p} + j\boldsymbol{q} \tag{2.8a}$$

$$\text{Three-phase} \rightarrow \underline{\boldsymbol{\mathcal{S}}} = \mathrm{diag}\left(\underline{\boldsymbol{\mathcal{U}}}\right) \underline{\boldsymbol{\mathcal{I}}}^* = \boldsymbol{\mathcal{P}} + j\boldsymbol{\mathcal{Q}} \tag{2.8b}$$

Including the grid with impedance matrix (2.1) and (2.6), the apparent power (2.8a),(2.8b) can be reformulated as

$$\text{Balanced equivalent} \rightarrow \underline{\boldsymbol{s}} = \mathrm{diag}\left(\underline{\boldsymbol{Z}}\, \underline{\boldsymbol{i}} + \mathbf{1}\underline{u}_0\right) \underline{\boldsymbol{i}}^* \tag{2.9a}$$

$$\text{Three-phase} \rightarrow \underline{\boldsymbol{\mathcal{S}}} = \mathrm{diag}\left(\underline{\mathbb{Z}}_{\mathrm{abcn}} \underline{\boldsymbol{\mathcal{I}}} + \mathbf{1} \otimes \underline{\boldsymbol{\mathcal{U}}}_0\right) \underline{\boldsymbol{\mathcal{I}}}^* \tag{2.9b}$$

In order to determine the generator power necessary to supply the loads within allowed limits of the power grid, the load and generator behavior can be described by equivalent models. The main models used in the literature are [Ker12]:

- **Constant impedance bus (Z)** – linear relationship between complex nodal voltage and current.

$$\underline{i} = -\frac{\underline{u}}{\underline{z}}$$

- **Constant current bus (I)** – fixed complex current injection no voltage dependency

$$\underline{i} = \underline{i}^{\text{const}} = \text{const.}$$

- **Constant power bus (P)** – fixed apparent power injection with nonlinear voltage dependency.

$$\underline{i} = \frac{\underline{s}}{\underline{u}}$$

- **Combination (ZIP)** – this load is a combination of all three load types, usually with scaling factors $\alpha_{\mathrm{Z}}, \alpha_{\mathrm{I}}, \alpha_{\mathrm{P}}$.

$$\underline{i} = -\alpha_{\mathrm{Z}}\frac{\underline{u}}{\underline{z}} + \alpha_{\mathrm{I}}\underline{i}^{\text{const}} + \alpha_{\mathrm{P}}\frac{\underline{s}}{\underline{u}}$$

- **Generator bus (PU)** – in this bus type the active power as well as the voltage magnitude $\|\underline{u}\|$ are assumed to be fixed. Reactive power is variable (excitation control).

PV units could be modeled as a PU bus type if the individual units are assumed to possess a feedback controller that actuates the reactive power injection automatically, so that the voltage magnitude would stay constant. In this work the constant power bus model is used to describe the PV and the demand in the nodes of the LV network.

Substation transformer model

The transformer at the substation connects the low voltage with the medium voltage grid. A common vector group for German 20/0.4 kV substation transformers is Dyn5 [PHS05]. Reading that the high voltage side is connected in delta and the low voltage side in wye with a connect-able star-point as depicted in Figure 2.4. Further it is assumed that the transformed MV voltage of the transformer is constant $\underline{\boldsymbol{\mathcal{U}}}_0 = \left[\underline{u}_0^{\mathrm{a}}, \underline{u}_0^{\mathrm{b}}, \underline{u}_0^{\mathrm{c}}\right]^{\mathrm{T}}$ and can be represented as an ideal voltage source on the low voltage side.

The impedance of the PEN connection is neglected and the transformer is represented by its positive-sequence short circuit impedance $\underline{z}_{\mathrm{k}}^{\mathrm{pos}}$ mapped onto the phases $\{\underline{z}_{\mathrm{k}}^{\mathrm{a}}, \underline{z}_{\mathrm{k}}^{\mathrm{b}}, \underline{z}_{\mathrm{k}}^{\mathrm{c}}\}$ transformed to the low voltage side. Mutual and stray inductance between the phases as well as parasitic capacitance of the coils are neglected. Furthermore, it is assumed that the voltage of the PEN conductor connected to the substation transformer is equal to $\underline{u}_0^{\mathrm{n}} \approx 0\,\mathrm{V}$.

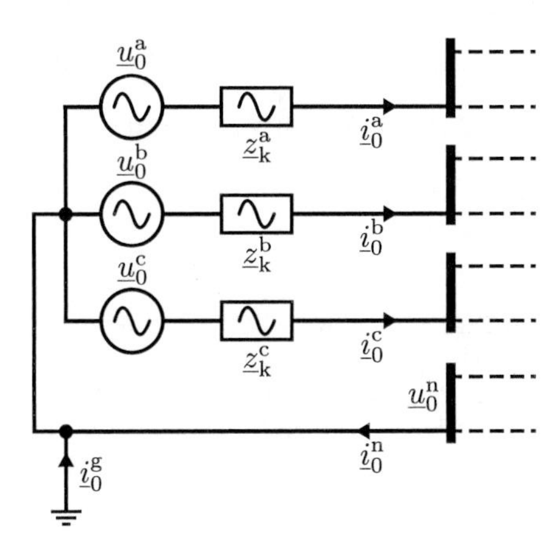

Figure 2.4: Equivalent model of the substation transformer with slack bus voltage

2.3 State-of-the-art solution to the distribution power flow problem

The power flow equations

Taking into consideration (2.1) and (2.8a) and rewriting them in terms of real and imaginary part, the power flow equations for active and reactive power of each node ν can be formulated as

$$
\begin{aligned}
-p_\nu + 2\Re\{\underline{u}_\nu\}^2\Re\{\underline{y}_{\nu,\nu}^{\Sigma}\} + \sum_{\mu=1,\mu\neq\nu}^{N_{+0}} \Re\{\underline{u}_\nu\}\Big(\Re\{\underline{u}_\mu\}\Re\{\underline{y}_{\nu,\mu}^{\Sigma}\} - \Im\{\underline{u}_\mu\}\Im\{\underline{y}_{\nu,\mu}^{\Sigma}\}\Big) + \\
\Im\{\underline{u}_\nu\}\Big(\Im\{\underline{u}_\mu\}\Re\{\underline{y}_{\nu,\mu}^{\Sigma}\} + \Re\{\underline{u}_\mu\}\Im\{\underline{y}_{\nu,\mu}^{\Sigma}\}\Big) = 0
\end{aligned}
\tag{2.10}
$$

$$\begin{aligned} -q_\nu + 2\Im\{\underline{u}_\nu\}^2\Im\{\underline{y}^{\Sigma}_{\nu,\nu}\} + \sum_{\mu=1,\mu\neq\nu}^{N_{+0}} \Re\{\underline{u}_\nu\}\left(\Re\{\underline{u}_\mu\}\Re\{\underline{y}^{\Sigma}_{\nu,\mu}\} + \Im\{\underline{u}_\mu\}\Im\{\underline{y}^{\Sigma}_{\nu,\mu}\}\right) + \\ \Im\{\underline{u}_\nu\}\left(\Im\{\underline{u}_\mu\}\Re\{\underline{y}^{\Sigma}_{\nu,\mu}\} - \Re\{\underline{u}_\mu\}\Im\{\underline{y}^{\Sigma}_{\nu,\mu}\}\right) = 0 \end{aligned} \tag{2.11}$$

where $\underline{y}^{\Sigma}_{\nu,\nu}$ denote the entries of $\underline{\boldsymbol{Y}}$ in Equation (2.1). If these equation are aligned properly and stacked for every node in the grid, following compact form can be found.

$$\boldsymbol{g}(\boldsymbol{x}) = \boldsymbol{0}, \quad \boldsymbol{x} = \left[\Re\{\underline{\boldsymbol{u}}\}^{+0}, \Im\{\underline{\boldsymbol{u}}\}^{+0}, \boldsymbol{p}^{+0}, \boldsymbol{q}^{+0}\right]^{\mathrm{T}} \tag{2.12}$$

With the vectors for the real part of the nodal voltage $\Re\{\underline{\boldsymbol{u}}\}^{+0} = [\Re\{\underline{u}_0\}, \cdots, \Re\{\underline{u}_N\}]^{\mathrm{T}}$, imaginary part $\Im\{\underline{\boldsymbol{u}}\}^{+0} = [\Im\{\underline{u}_0\}, \cdots, \Im\{\underline{u}_N\}]^{\mathrm{T}}$, active power $\boldsymbol{p}^{+0} = [p_0, \cdots, p_N]^{\mathrm{T}}$ and reactive power $\boldsymbol{q}^{+0} = [q_0, \cdots, q_N]^{\mathrm{T}}$. The notation +0 indicates that the root node, which could be the slack bus, is included in the formulation.

2.3.1 Newton-Raphson method

The Newton-Raphson (NR) method solves the power flow (2.12) by approximating the nonlinear set of equations with a first order Taylor series at a current solution $\boldsymbol{x}^j$

$$\boldsymbol{g}(\boldsymbol{x}) \approx \boldsymbol{g}\left(\boldsymbol{x}^j\right) + \nabla^{\mathrm{T}}\boldsymbol{g}\left(\boldsymbol{x}^j\right)\Delta\boldsymbol{x}^j = \boldsymbol{0}.$$

To find the Newton direction $\Delta\boldsymbol{x}^j$ the Jacobian matrix $\nabla^{\mathrm{T}}\boldsymbol{g}\left(\boldsymbol{x}^j\right)$ which includes the partial derivatives, must be inverted. The new iterate $\boldsymbol{x}^{j+1}$ is computed by adding the Newton direction with a scalar weighting to the last iterate.

$$\begin{aligned} -\left[\nabla^{\mathrm{T}}\boldsymbol{g}\left(\boldsymbol{x}^j\right)\right]^{-1}\boldsymbol{g}\left(\boldsymbol{x}^j\right) &= \Delta\boldsymbol{x}^j \\ \boldsymbol{x}^{j+1} &= \boldsymbol{x}^j + \alpha\Delta\boldsymbol{x}^j \end{aligned} \tag{2.13}$$

Choosing the step size α of the Newton direction is crucial and can be efficiently implemented with backtracking line search [BV04]. The inversion of the Jacobi matrix is the most time-consuming part in the algorithm. Decoupling methods that intend to make the Jacobi matrix sparse [ZC95], can reduce the computation time for the inversion. A clear drawback of the NR method for low voltage distribution grids is the inability to handle the high R/X ratios. These high ratio leads to an ill conditioning of the Jacobi matrix, because of the partial derivatives become very small. Consequently convergence problems are the outcome [SHSL88].

2.3.2 Forward-Backward-Sweep method

To overcome the computational issues of the NR, the Forward-Backward-Sweep (FBS) algorithm was developed for radial distribution grids. From an algorithmic perspective the idea of the FBS method is to sweep from the terminal nodes to the root node computing the branch currents with the voltage iterate for every node. After that the nodal voltages are updated with the previous calculated branch currents sweeping from the root node to the terminal nodes.

In the forward sweep the current injections for every node are calculated for a constant power load

$$\underline{i}_{\nu}^{j} = \frac{\underline{s}_{\nu}^{*}}{\left(\underline{u}_{\nu}^{j-1}\right)^{*}} \quad \forall \nu \in \mathcal{N}. \tag{2.14}$$

And in the backward sweep the nodal voltage are updated with the calculated currents

$$\underline{\boldsymbol{u}}^{j} = \underline{\boldsymbol{Z}}\,\underline{\boldsymbol{i}}^{j} + \mathbf{1}\underline{u}_{0}. \tag{2.15}$$

with j again being the iteration index. Both (2.14) and (2.15) are repeatedly evaluated in a sequence till a convergence is reached. As a convergence criterion the apparent power mismatch $\left\| \left(\underline{\boldsymbol{i}}^{j}\right)^{*} \underline{\boldsymbol{u}}^{j-1} \right\| - \|\underline{\boldsymbol{s}}\| \leq \delta_{s}$ is usually chosen to terminate the algorithm for the constant power load. From (2.14) and (2.15) it is obvious that there are no expensive matrix inversion involved in the calculation process. Even though experience shows that the FBS needs more iterations than the NR method to converge to a solution, it still outperforms the NR method in terms of computation time [Zim95]. This advantage comes at the expanse of PU-bus models, which cannot simply be integrated in the algorithm.

2.3.3 General relationship of nodal voltages and the power flow

For the investigation in this thesis the voltage and apparent power interdependence is one of the key factors. To describe this behavior in more detail, a two node example with one branch (left), and the corresponding complex voltage phasor diagram (right) is constructed, as depicted in Figure 2.5.

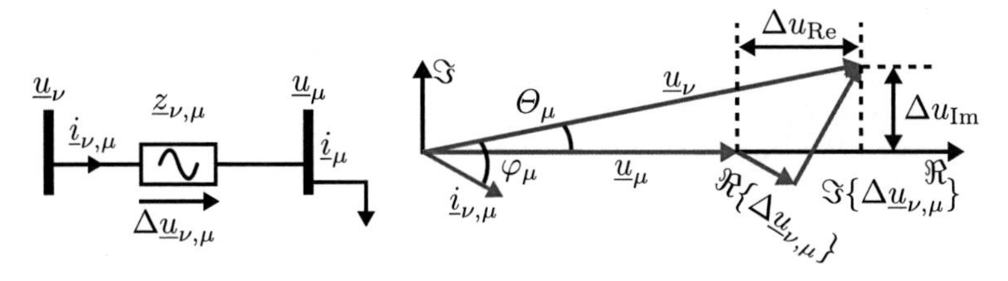

Figure 2.5: Example phasor diagram for a single branch with two nodes

For the calculation following assumptions must be introduced: the phase angle $\measuredangle \underline{u}_\nu = 0$ and voltage magnitude $\|\underline{u}_\nu\| = 1\,\text{p.u.}$ of the first node must be fixed and known. The respective voltage angle of the second node is $\measuredangle \underline{u}_\mu = \Theta_\mu$. Now let the phase angle for $\underline{s}_\mu$ be φ_μ and the impedance of the branch have a X/R ratio of 0.5. Following relationship for the voltages can now be found

$$\|\underline{u}_\nu\|^2 = \left(\|\underline{u}_\mu\| + \Delta u_{\text{Im}}\right)^2 + \Delta u_{\text{Re}}^2 \tag{2.16}$$

If the voltage drop Δu_{Im} and Δu_{Re} in Equation 2.16 are expressed in terms of the branch current $\underline{i}_{\nu,\mu}$ and phase angle φ_μ and then substituted by the respective active and reactive power p_μ,q_μ following expression is found

$$\begin{aligned}
\|\underline{u}_\nu\|^2 &= \left(\|\underline{u}_\mu\| + \frac{p_\mu r_{\nu,\mu} + q_\mu x_{\nu,\mu}}{\|\underline{u}_\mu\|}\right)^2 + \left(\frac{p_\mu x_{\nu,\mu} - q_\mu r_{\nu,\mu}}{\|\underline{u}_\mu\|}\right)^2 \\
\|\underline{u}_\nu\|^2 &= \|\underline{u}_\mu\|^2 + 2\left(p_\mu r_{\nu,\mu} + q_\mu x_{\nu,\mu}\right) + \left(\frac{p_\mu r_{\nu,\mu} + q_\mu x_{\nu,\mu}}{\|\underline{u}_\mu\|}\right)^2 + \left(\frac{p_\mu x_{\nu,\mu} - q_\mu r_{\nu,\mu}}{\|\underline{u}_\mu\|}\right)^2 .
\end{aligned} \tag{2.17}$$

Using φ_μ as a parameter, Equation 2.17 can now be rewritten with $q_\mu = p_\mu \tan(\varphi_\mu)$ being the only variable. Furthermore, $\|\underline{u}_\nu\| = 1\,\text{p.u.}$ is substituted into the resulting equation and both sides are multiplied with $\|\underline{u}_\mu\|^2$. This leads to the following expression

$$\begin{aligned}
0 = \|\underline{u}_\mu\|^4 + 2\|\underline{u}_\mu\|^2\left(p_\mu r_{\nu,\mu} + p_\mu \tan(\varphi_\mu) x_{\nu,\mu} - \frac{1}{2}\right) + \left(p_\mu r_{\nu,\mu} + p_\mu \tan(\varphi_\mu) x_{\nu,\mu}\right)^2 + \\
\left(p_\mu x_{\nu,\mu} - p_\mu \tan(\varphi_\mu) r_{\nu,\mu}\right)^2
\end{aligned} \tag{2.18}$$

Equation 2.18 is now further used to introduce the concept of voltage stability as the expression builds the basis to calculate the voltage curve.

Voltage curve and feasible solutions to the power flow

As shown in the derivation, if there is an apparent power exchange between a node and the grid, the voltage at the particular node will change according to the power flow equations (2.11) and (2.10). For the simple example depicted in Figure 2.5 this relationship can be calculated using (2.18). A qualitative depiction of the obtained results based on the simplified circuit is shown in Figure 2.6. Where $\|\underline{u}_0\|$ is the nominal value on the y-axis and p_μ the active power drawn from the grid in node μ on the x-axis. The active power is increased from zero up to a certain value $p_\mu = \{0, \cdots, p_{\mu,\text{crit.}}\}$ and the relation between active and reactive power described by $\tan(\varphi_\mu)$ is kept constant. At a certain loading situation $p_{\mu,\text{crit.}}$ at the second node, a further increase of power will lead to an unstable operating point. In general, it is not guaranteed that power flow algorithms find a solution if the resulting voltage is close to this point. One physical reason for this is the so-called "voltage stability" phenomenon [Lar00]. Power lines are simply only able to transmit a finite amount of power. The point where this phenomenon happens is often referred to as the critical voltage $||\underline{u}_{\text{crit}}||$. Most power flow approximations make use of

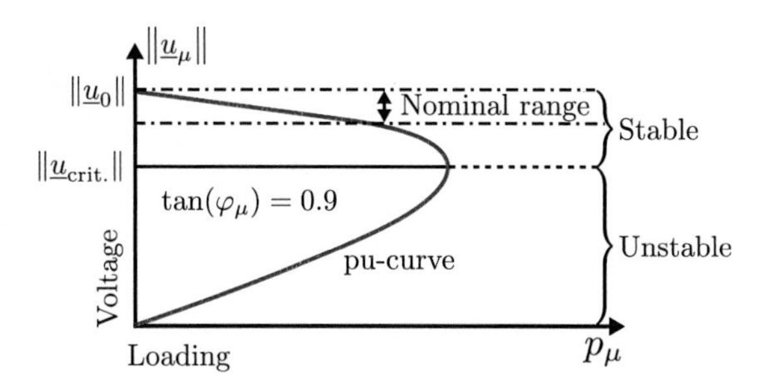

Figure 2.6: pu-curve for a simple two-node example

the linear behavior of the voltage described by the nose curve in the nominal range. A brief overview of these methods will be given in the next section.

2.4 Efficient power flow approximations

The power flow equations are non-linear and non-convex [COC12]. That is both true for the polar and the rectangular representation, which makes them a difficult constraint for the inclusion in an optimal power flow or dynamic optimal power flow problem (1.1a). In order to achieve better properties for the optimization problem, approximations of the power flow can be used instead. Most approximations found in the literature have a linear convex character. They use simplifications and knowledge about the power system and as such, are only valid for these special situations. Despite this, radial distribution systems and especially the development of acquisition and communication technologies have again triggered extensive research for approximations with additional flexibility. Especially, the ability to adapt to an operating point, which is not in the proximity of nominal values has gained significant interest. Further, in the section, selected approximations from literature a separated into three main groups and discussed in detail.

Group 1: approximations based on nominal values computed off-line

The idea behind these approaches is to develop and off-line approximation based on the no-load solution. This approximation is then used on-line to calculate the power flow. Here, the DC power flow [KCK13], linear distribution power flow (LinDistFlow) [KZGB16] and the linear coupled power flow (LCPF) [DGC15] are the most notable of these formulations. They have a long history [BW89, Woo96], and proven very usefull. A newer approach presented in [BZ15] first proves the existence of a power flow solution in a radial network and develops an approximate formulation by neglecting the non-linear voltage dependency of the constant power load model. This way, they can formulate an explicit linear relationship between apparent power and voltage. However, the approximation only produces satisfactory results if the nodal voltages are close to the nominal value. The ability to adapt to an operating point, which is not in the proximity of nominal values has gained significant interest. To give a brief overview of the most notable methods in the literature, these are further discussed.

Group 2: approximations with piecewise constant apparent power, some computation on-line

This group of approximations calculates a general formulation off-line and adds a dynamic part to the equation which changes on-line. An approximation developed in [BCCC11] is based on a second-order Taylor series expansion for an infinite slack bus voltage. Apparent power is seen a piecewise constant input and state. The power flow model relates the nodal voltage and apparent power by a first-order dynamic equation. This dynamic part is used to bridge the transition between changes of apparent power. An extension of this formulation for the more general ZIP-load model, but at the same time neglecting the dynamics, has been proposed in [Gar16]. Because constant impedance and constant current are both linear models, only the non-linear constant power bus model is approximated. The obtained linear approximation is further extended for three-phase power grids. For power flow calculations, these methods are beneficial and accurate. However, a clear drawback of these methods is that they can only be used for an optimal power flow (OPF) when the apparent power of the load undergoes small changes. Otherwise, the assumption of piecewise constant values is not valid.

Group 3: approximations with an on-line generation of the approximation

Often it is desired to dispatch flexible loads and distributed generation in an optimal power flow or dynamic optimal power flow. Both approximations [BCCC11] and [Gar16] are not applicable as they are bilinear in apparent power and voltage. However, in

[Gar15], the authors show how to solve an optimal power flow with this approximation. They especially address how to handle the bilinear character of the power flow approximation. An alternative approach from [BD15] implicitly linearize the power flow manifold by a first-order Taylor approximation in order to generate a tangent plane at a certain operating point. Depending on the curvature of the manifold this tangent plane approximation performs well in a certain region around its development point. In order to acquire this development point, two possible solution strategies can be chosen. Either a power flow must be solved in every instant the model is generated, or PMU units has to be dispatched to all nodes in the grid. A phase angle for every node is mandatory to calculate the orientation of the tangent plane. Neglecting the phase angles resolves in the LCPF. Another promising approach using the rectangular formulation of the power flow is presented in [BWA+17]. Rather, then performing a first-order Taylor approximation, the authors perform a fixed point linearization. This again can be either achieved by solving a full power flow up-front or use PMUs to receive magnitude and phase angles to calculate the fixed point. Because this method is compelling, other authors have also found adapted versions with certain simplifications as in [For17]. By exchanging the voltages needed to calculate the power flow solution by real measurements from smart-meters, a reduced formulation can be derived. However, the simplification only works if the phase angle of the nodal voltage is small, which, on the other hand, means that reactive power exchange with the grid must be small.

Summary of the approximation methods

In Table 2.1, the discussed approaches which can directly be used for optimal power flow or dynamic optimal power flow calculations are compared. This excludes the second group of algorithms, which have been introduced before. The table directly shows that neglecting losses results in low computational demands, as these methods do not need on-line reevaluation. On the other hand, the algorithms that need a recalculation of some matrices during run-time, tend to increase the accuracy but have a higher computational demand. At the end of the accuracy spectrum are methods, that either need full power flow calculations or PMU readings collected on-line to work properly.

2.4.1 Linear interpolation power flow

This section introduces the linear interpolation power flow (LIPF) which was developed in this thesis. It has similar accuracy features as the fixed point linearization introduced in [BWA+17], but does not need phase information at every node. Instead of solving a power flow or measuring with PMUs, it produces this information by utilizing the communication infrastructure in the grid collecting measurements from smart-meters

Table 2.1: Comparison of the different most significant power flow approximations found in literature for the use in low voltage grids

On-line evaluation	Accuracy	Type	Losses	PMU, PF	Reference
×	low	Implicit	×	×	[KCK13]
×	low	Explicit	×	×	[BZ15]
×	mid	Explicit	×	×	[KZGB16]
×	mid	Implicit	×	×	[DGC15]
✓	mid	Explicit	✓	×	[For17]
✓	high	Implicit	✓	✓	[BD15]
✓	high	Explicit	✓	✓	[BWA+17]

connected to the nodes. Voltage magnitude and apparent power are then used to calculate an approximate solution to a fixed point iteration of the FBS method (2.14)+(2.15). This solution, which can be obtained by a single matrix inversion, then serves as a basis for further calculations. The power flow approximation interpolates the nonlinear manifold between the no-load solution at nominal voltage and the approximated fixed point iteration at the current operating point $\|\overline{\underline{u}}_\mu\|, \overline{p}_\mu, \overline{q}_\mu$. A comparison of the three groups of approximations introduced in the preceding section is shown together with the upper part of the pu-curve from Figure 2.6 in Figure 2.7.

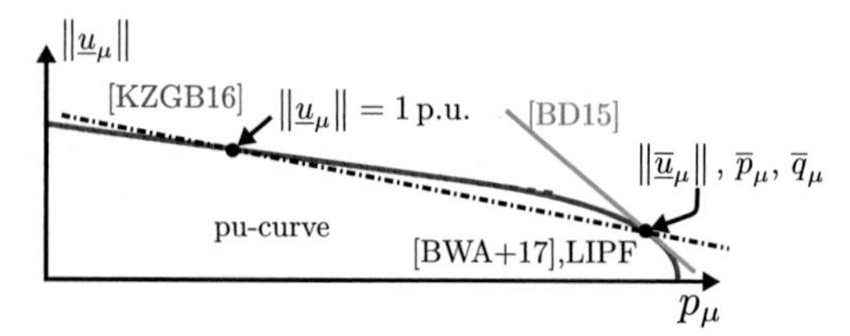

Figure 2.7: Representation of the three main approximation approaches reviewed in this thesis. LinDistFlow [KZGB16] (red-dashed line), methods using a tangent plane [BD15] (green-solid line), fixed point linearization [BWA+17], LIPF (black-dash-dotted line), the non-linear solution with FBS (blue-solid line)

The use of on-line approximations in grid operation and control

The LIPF was specifically designed for the integration into a receding horizon control solving an optimal power flow. By using the measured feedback from the grid to set the parameters of the approximation, it is always adapted to the current operating point. Figure 2.8 depicts the work-flow and interaction of the approximation with grid,

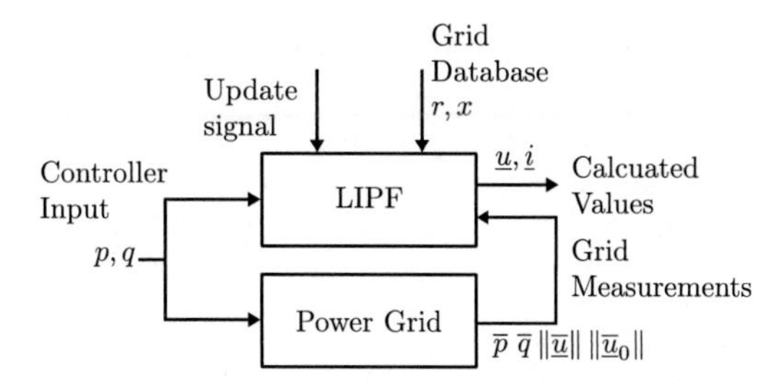

Figure 2.8: Interaction between the LIPF, the controller and the power grid

controller and databases.

Update signal

The recalculation of the power flow approximation uses an update signal as a trigger. The need for this signal can be calculated based on different criteria like accuracy, fast changes in voltages, or an on-line change of parameters. As long as this trigger is false, there is no need to recalculate the approximation. For the remainder of this thesis, there is no specific update signal evaluated. The LIPF is recalculated after every control cycle of the MPC.

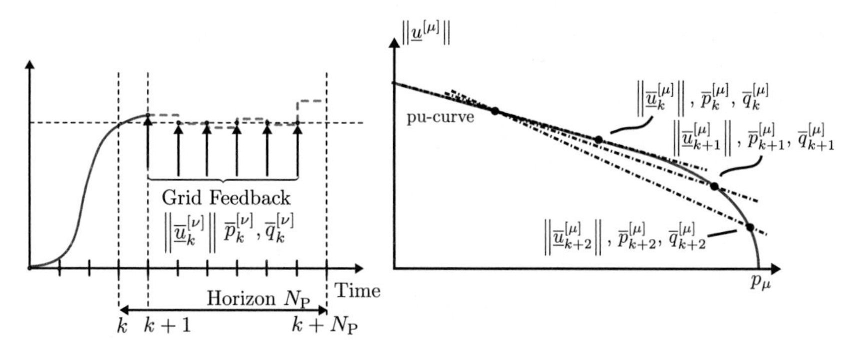

Figure 2.9: Updates and recalculation of the LIPF, with the MPC trajectory on the left (see also Figure 1.8) and the exemplary operating points for the respective steps $\{k, k+1, k+2\}$ on the right

Controller input

Calculated power outputs from the supervisory DMPC serve as an input to the LIPF to calculate the complex currents and voltages of the nodes in the grid in the balanced case. The three-phase approximation interacts with the balancing controller.

Grid database
As the power flow approximation needs information about the grid infrastructure and the line parameters, it has some interface to a database of a grid operation center. Additionally, it can receive parameters in case the controller make use of on-line parameter estimation.

Grid measurements
The distributed controllers collect and exchange measurements from different nodes in the grid. These measurements serve as an additional input to the LIPF. Each local controller has its own set of known grid measurements and interprets them. The values are only integrated if the update signal is **TRUE**.

Calculated power flow
Resulting from the linear power flow approximation are the complex currents and voltages. These are used in the distributed control to predict the future behavior of the grid. Because the MPC repeats the prediction and calculation of control inputs cyclically, as explained in Section 1.2.2, the LIPF can be updated with grid measurements in every one of these instances. Figure 2.9 illustrates this mechanism for different discrete-time instances $k+l$ with $l \in \{0, \cdots, N_\mathrm{P}\}$.

In the following section, first, the balanced single-phase representation of the LIPF is introduced and then extended to the three-phase version. Afterward, the decomposition into real and imaginary parts is introduced and the transformation in sequence-frame with symmetrical components is presented.

Balanced single-phase formulation

It is assumed that consumers and generation units in low voltage grids can be modeled as a constant power bus model leading to a nodal current injection of the following form

$$\underline{i}_\nu = \frac{\underline{s}_\nu^*}{\underline{u}_\nu^*}. \tag{2.19}$$

By adopting the nodal voltage equation (2.2), $\underline{u}_\nu$ can be reformulated as a function of all nodal currents $\underline{\boldsymbol{i}}$ and the slack bus voltage $\underline{u}_0$, i.e.

$$\underline{u}_\nu = \mathbf{1}_\nu^\mathrm{T} \underline{\boldsymbol{Z}}\, \underline{\boldsymbol{i}} + \underline{u}_0 \tag{2.20}$$

where $\mathbf{1}_\nu = [0, \ldots, 1, \ldots 0]^\mathrm{T}$ is a vector with entry ν equal to 1 and other all entries equal to zero. It maps the corresponding voltage drops across the power lines from the slack bus to node ν. The number of elements in $\mathbf{1}_\nu$ is N

Substituting (2.20) into (2.19) with $\underline{\epsilon}_0 = 1/\underline{u}_0^*$ leads to

$$-\underline{i}_\nu - \underline{i}_\nu\,\underline{\epsilon}_0\,\mathbf{1}_\nu^{\mathrm{T}}\underline{\boldsymbol{Z}}^*\underline{\boldsymbol{i}}^* + \underline{\epsilon}_0\,\underline{s}_\nu^* = 0. \tag{2.21}$$

Assuming that nodal apparent power and voltage are measurable, it is possible to recover a valid power flow solution $\overline{\underline{u}}_\nu$,$\overline{\underline{s}}_\nu \forall\nu \in \mathcal{N}$ that satisfies the power flow equations of the network. This means that the power grid itself solves the power flow, which is a reasonable assumption [GL16]. The nodal current in (2.21) which is the source of the nonlinearity, is substituted with the measured power flow solution as $\overline{\underline{s}}_\nu^*/\overline{\underline{u}}_\nu^* = \overline{\underline{\epsilon}}_\nu\overline{\underline{s}}_\nu^* = \overline{\underline{i}}_\nu \approx \underline{i}_\nu$. The resulting formulation can be written as

$$-\underline{i}_\nu - \overline{\underline{\epsilon}}_\nu\,\overline{\underline{s}}_\nu^*\underline{\epsilon}_0\mathbf{1}_\nu^{\mathrm{T}}\underline{\boldsymbol{Z}}^*\underline{\boldsymbol{i}}^* + \underline{\epsilon}_0\,\underline{s}_\nu^* = 0. \tag{2.22}$$

By stacking up (2.22) for all nodes ν the approximated power flow equations for the grid can be found i.e.

$$-\underline{\boldsymbol{i}} - \overline{\underline{\boldsymbol{E}}}\,\overline{\underline{\boldsymbol{S}}}^*\underline{\boldsymbol{E}}_0\underline{\boldsymbol{Z}}^*\underline{\boldsymbol{i}}^* + \underline{\boldsymbol{E}}_0\,\underline{\boldsymbol{s}}^* = \mathbf{0}. \tag{2.23}$$

with $\overline{\underline{\boldsymbol{S}}} = \mathrm{diag}\,(\overline{\underline{s}}_1,\ldots,\overline{\underline{s}}_N)$, $\overline{\underline{\boldsymbol{E}}} = \mathrm{diag}\,(\overline{\underline{\epsilon}}_1,\ldots,\overline{\underline{\epsilon}}_N)$,$\underline{\boldsymbol{E}}_0 = \mathrm{diag}\,(\underline{\epsilon}_0,\ldots,\underline{\epsilon}_0)$, $\underline{\boldsymbol{s}} = [\underline{s}_1,\ldots,\underline{s}_N]^{\mathrm{T}}$ and $\underline{\boldsymbol{i}} = [\underline{i}_1,\ldots,\underline{i}_N]^{\mathrm{T}}$. Note that $\mathrm{diag}\,(a_1,\ldots,a_N)$ is a matrix with only $[a_1,\ldots,a_N]$ as diagonal elements.

Extension to three-phase four-wire systems

In the same manner as in (2.20) the three-phase voltages $\underline{u}_\nu^{\mathrm{a}}$, $\underline{u}_\nu^{\mathrm{b}}$, $\underline{u}_\nu^{\mathrm{c}}$ are reformulated as a function of all nodal currents $\underline{i}_\nu^{\mathrm{a}}$, $\underline{i}_\nu^{\mathrm{b}}$, $\underline{i}_\nu^{\mathrm{c}}$ and the slack bus voltage, i.e.

$$\begin{pmatrix}\underline{u}_\nu^{\mathrm{a}}\\ \underline{u}_\nu^{\mathrm{b}}\\ \underline{u}_\nu^{\mathrm{c}}\end{pmatrix} = \begin{pmatrix}\mathbb{1}_\nu^{\mathrm{a,T}}\\ \mathbb{1}_\nu^{\mathrm{b,T}}\\ \mathbb{1}_\nu^{\mathrm{c,T}}\end{pmatrix}\underline{\mathbb{Z}}_{\mathrm{abcn}}\underline{\boldsymbol{\mathcal{I}}}_\nu + \begin{pmatrix}1\\ \underline{m}\\ \underline{m}^2\end{pmatrix}\underline{u}_0 \tag{2.24}$$

where $\mathbb{1}_\nu^{\mathrm{a}} = [0,\ldots,1,\ldots 0]^T$ is now a vector with the entry at node ν phase $\mathbf{a}$ equal to 1 and all other entries equal to zero. It maps the corresponding voltage drops in every phase from the slack bus to node ν. The number of elements in $\mathbb{1}_\nu^{\mathrm{a}}, \mathbb{1}_\nu^{\mathrm{b}}, \mathbb{1}_\nu^{\mathrm{c}}$ is $3N$.

By measuring a solution to the three-phase power flow as in (2.22) the following equation can be found

$$-\underline{i}_\nu^\xi - \overline{\underline{\epsilon}}_\nu^\xi\left(\overline{\underline{s}}_\nu^\xi\right)^*\underline{\epsilon}_0^\xi\,\mathbb{1}_\nu^{\xi,\mathrm{T}}\underline{\mathbb{Z}}_{\mathrm{abcn}}^*\underline{\boldsymbol{\mathcal{I}}}_\nu^* + \underline{\epsilon}_0^\xi\left(\underline{s}_\nu^\xi\right)^* = 0. \tag{2.25}$$

Again by stacking up (2.25), but now for all nodes ν and phases ordered $\xi = \{\mathrm{a,b,c}\}$

$$-\underline{\boldsymbol{\mathcal{I}}} - \overline{\underline{\mathbb{E}}}\,\overline{\underline{\mathbb{S}}}^*\underline{\mathbb{E}}_0\underline{\mathbb{Z}}_{\mathrm{abcn}}^*\underline{\boldsymbol{\mathcal{I}}}^* + \underline{\mathbb{E}}_0\,\underline{\boldsymbol{S}}^* = \mathbf{0} \tag{2.26}$$

with the measured nodal voltage vector $\overline{\mathbb{E}} = \text{diag}\left(\overline{\underline{\epsilon}}_1^{\text{a}}, \overline{\underline{\epsilon}}_1^{\text{b}}, \overline{\underline{\epsilon}}_1^{\text{c}}, \ldots \overline{\underline{\epsilon}}_N^{\text{a}}, \overline{\underline{\epsilon}}_N^{\text{b}}, \overline{\underline{\epsilon}}_N^{\text{c}}\right)$ the inverse slack bus voltage in all three phases $\mathbb{E}_0 = \text{diag}\left(\underline{\epsilon}_0^{\text{a}}, \underline{\epsilon}_0^{\text{b}}, \underline{\epsilon}_0^{\text{c}}, \ldots \underline{\epsilon}_0^{\text{a}}, \underline{\epsilon}_0^{\text{b}}, \underline{\epsilon}_0^{\text{c}}\right)$, the matrix of measured nodal apparent power $\overline{\underline{\mathbb{S}}} = \text{diag}\left(\overline{\underline{s}}_1^{\text{a}}, \overline{\underline{s}}_1^{\text{b}}, \overline{\underline{s}}_1^{\text{c}}, \ldots \overline{\underline{s}}_N^{\text{a}}, \overline{\underline{s}}_N^{\text{b}}, \overline{\underline{s}}_N^{\text{c}}\right)$, $\underline{\boldsymbol{S}} = \left[\underline{s}_1^{\text{a}}, \underline{s}_1^{\text{b}}, \underline{s}_1^{\text{c}}, \ldots, \underline{s}_N^{\text{a}}, \underline{s}_N^{\text{b}}, \underline{s}_N^{\text{c}}\right]^{\text{T}}$. Both (2.23) and (2.26) are linear complex valued functions in $\underline{\boldsymbol{i}}$ and $\underline{\boldsymbol{s}}$ for the single-phase equivalent and $\underline{\boldsymbol{I}}$ and $\underline{\boldsymbol{S}}$ for the three-phase four-wire case.

Inclusion of shunt impedances

To use the approximated power flow method for the estimation of medium voltage grids beyond the substation transformer shunt impedance can no longer be neglected. In particular, the capacitance of the medium voltage cables needs to be considered. This can be achieved by simply adding the shunt impedances introduced in Figure 2.3. The overall equation leads to (2.27)

$$-\underline{\boldsymbol{I}} - \overline{\mathbb{E}}\,\overline{\underline{\mathbb{S}}}^{*}\mathbb{E}_0\underline{\mathbb{Z}}_{\text{abcn}}^{*}\underline{\boldsymbol{I}}^{*} + \mathbb{E}_0\,\underline{\boldsymbol{S}}^{*} + \underline{\mathbb{Z}}_{\text{sh}}^{-1}\left(\underline{\mathbb{Z}}_{\text{abcn}}\underline{\boldsymbol{I}} + \mathbb{E}_0^{-1}\mathbb{1}\right) = \mathbf{0} \tag{2.27}$$

where $\mathbb{1}$ is now a vector with all elements equal to 1 with size $3N$.

Representation of unbalanced three-phase systems with symmetrical components

So far the approximated power flow is able to represent unbalanced power grids with three phases and a neutral conductor. The complex line to neutral voltages that can be locally measured at each node in a grid with the numbering {a, b, c} can be transformed into three sequence-components {pos, neg, zero}. Through this transformation, which can be achieved with the following linear relationship [CFO03], the unbalance in the grid can be further investigated.

$$\begin{pmatrix} \underline{u}_\nu^{\text{pos}} \\ \underline{u}_\nu^{\text{neg}} \\ \underline{u}_\nu^{\text{zero}} \end{pmatrix} = \underbrace{\frac{1}{3}\begin{pmatrix} 1 & \underline{m} & \underline{m}^2 \\ 1 & \underline{m}^2 & \underline{m} \\ 1 & 1 & 1 \end{pmatrix}}_{\underline{\boldsymbol{T}}_{\text{pnz}}} \begin{pmatrix} \underline{u}_\nu^{\text{a}} \\ \underline{u}_\nu^{\text{b}} \\ \underline{u}_\nu^{\text{c}} \end{pmatrix}. \tag{2.28}$$

Through the transformation (2.28) the unsymmetrical phase voltages $\underline{u}_\nu^{\text{a}}, \underline{u}_\nu^{\text{b}}, \underline{u}_\nu^{\text{c}}$ can be represented in three symmetrical voltages $\underline{u}_\nu^{\text{pos}}, \underline{u}_\nu^{\text{neg}}, \underline{u}_\nu^{\text{zero}}$ as depicted in Figure 2.10. The grey arrows depict a balanced reference voltage.

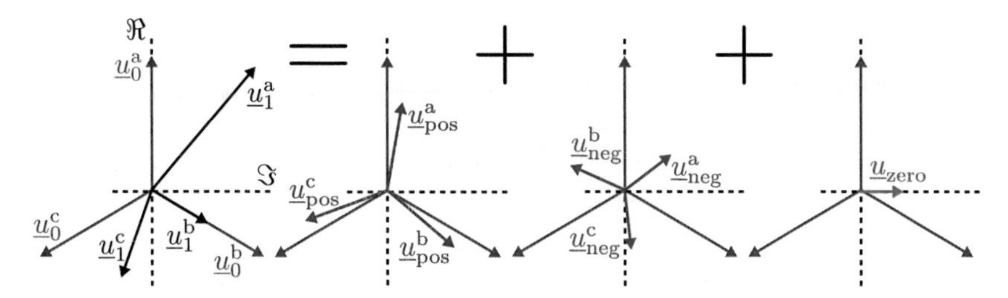

Figure 2.10: Transformation of three unbalanced phase-components {a, b, c} into three balanced sequence-components {positive (pos), negative (neg), zero}

2.4.2 Real valued decomposition

Phase-frame based approximated power flow decomposition

In order to integrate conjugate and non-conjugate variables in one set of equations a real-valued formulation is derived by decomposing the complex vectors and matrices of (2.27). The variables are reordered to real and imaginary part, so they can be expressed in a more compact form, which is used further in this thesis. The individual matrices are decomposed in real and imaginary part. For the slack bus and measured nodal voltage, this leads to $\Re\{\underline{\mathbb{E}}_0\} = \mathbb{E}_0^{\mathrm{Re}}, \Im\{\underline{\mathbb{E}}_0\} = \mathbb{E}_0^{\mathrm{Im}}, \Im\{\overline{\underline{\mathbb{E}}}\} = \overline{\mathbb{E}}^{\mathrm{Im}}, \Re\{\overline{\underline{\mathbb{E}}}\} = \overline{\mathbb{E}}^{\mathrm{Re}}$. The grid impedance with the included neutral conductor and the shunt matrix the decomposition takes on the following from: $\Re\{\underline{\mathbb{Z}}_{\mathrm{abcn}}\} = \mathbb{R}$, $\Im\{\underline{\mathbb{Z}}_{\mathrm{abcn}}\} = \mathbb{X}$, $\Re\{\underline{\mathbb{Z}}_{\mathrm{sh}}^{-1}\} = \mathbb{G}_{\mathrm{sh}}$, $\Im\{\underline{\mathbb{Z}}_{\mathrm{sh}}^{-1}\} = \mathbb{B}_{\mathrm{sh}}$. The measured apparent power matrices are decomposed as $\Re\{\overline{\underline{\mathbb{S}}}\} = \overline{\mathbb{P}}$, $\Im\{\overline{\underline{\mathbb{S}}}\} = \overline{\mathbb{Q}}$ and the variable nodal power injections decomposes as $\Re\{\underline{\boldsymbol{\mathcal{S}}}\} = \boldsymbol{\mathcal{P}}$, $\Im\{\underline{\boldsymbol{\mathcal{S}}}\} = \boldsymbol{\mathcal{Q}}$, $\Re\{\underline{\boldsymbol{\mathcal{I}}}\} = \boldsymbol{\mathcal{I}}^{\mathrm{Re}}$, $\Im\{\underline{\boldsymbol{\mathcal{I}}}\} = \boldsymbol{\mathcal{I}}^{\mathrm{Im}}$.

To further simplify the expression $\mathbb{J} = \overline{\mathbb{Q}}\left(\mathbb{E}_0^{\mathrm{Im}}\mathbb{X} - \mathbb{E}_0^{\mathrm{Re}}\mathbb{R}\right) - \overline{\mathbb{P}}\left(\mathbb{E}_0^{\mathrm{Re}}\mathbb{X} + \mathbb{E}_0^{\mathrm{Im}}\mathbb{R}\right)$ and $\mathbb{V} = \overline{\mathbb{Q}}\left(\mathbb{E}_0^{\mathrm{Re}}\mathbb{X} + \mathbb{E}_0^{\mathrm{Im}}\mathbb{R}\right) + \overline{\mathbb{P}}\left(\mathbb{E}_0^{\mathrm{Im}}\mathbb{X} + \mathbb{E}_0^{\mathrm{Re}}\mathbb{R}\right)$ are introduced. Now the approximated power flow can be formulated as

$$\underbrace{\begin{pmatrix} -\boldsymbol{I}^{6N\times 6N} - \overline{\mathbb{E}}^{\mathrm{Im}}\mathbb{J} + \overline{\mathbb{E}}^{\mathrm{Re}}\mathbb{V} + \mathbb{G}_{\mathrm{sh}}\mathbb{R} - \mathbb{B}_{\mathrm{sh}}\mathbb{X} & -\overline{\mathbb{E}}^{\mathrm{Im}}\mathbb{V} - \overline{\mathbb{E}}^{\mathrm{Re}}\mathbb{J} - \mathbb{G}_{\mathrm{sh}}\mathbb{X} - \mathbb{B}_{\mathrm{sh}}\mathbb{R} \\ -\overline{\mathbb{E}}^{\mathrm{Im}}\mathbb{V} - \overline{\mathbb{E}}^{\mathrm{Re}}\mathbb{J} + \mathbb{B}_{\mathrm{sh}}\mathbb{R} + \mathbb{G}_{\mathrm{sh}}\mathbb{X} & -\boldsymbol{I}^{6N\times 6N} + \overline{\mathbb{E}}^{\mathrm{Im}}\mathbb{J} - \overline{\mathbb{E}}^{\mathrm{Re}}\mathbb{V} + \mathbb{G}_{\mathrm{sh}}\mathbb{R} - \mathbb{B}_{\mathrm{sh}}\mathbb{X} \end{pmatrix}}_{\mathbb{A}_{\mathrm{PF}}} \underbrace{\begin{pmatrix} \boldsymbol{\mathcal{I}}^{\mathrm{Re}} \\ \boldsymbol{\mathcal{I}}^{\mathrm{Im}} \end{pmatrix}}_{\boldsymbol{\mathcal{I}}}$$

$$+ \underbrace{\begin{pmatrix} \mathbb{E}_0^{\mathrm{Re}} & -\mathbb{E}_0^{\mathrm{Im}} \\ -\mathbb{E}_0^{\mathrm{Im}} & -\mathbb{E}_0^{\mathrm{Re}} \end{pmatrix}}_{\mathbb{B}_{\mathrm{PF}}} \underbrace{\begin{pmatrix} \boldsymbol{\mathcal{P}} \\ \boldsymbol{\mathcal{Q}} \end{pmatrix}}_{\boldsymbol{\mathcal{S}}} - \underbrace{\begin{pmatrix} \mathbb{U}_0^{\mathrm{Re}} & -\mathbb{U}_0^{\mathrm{Im}} \\ \mathbb{U}_0^{\mathrm{Im}} & \mathbb{U}_0^{\mathrm{Re}} \end{pmatrix} \begin{pmatrix} \mathbb{G}_{\mathrm{sh}} & -\mathbb{B}_{\mathrm{sh}} \\ \mathbb{B}_{\mathrm{sh}} & \mathbb{G}_{\mathrm{sh}} \end{pmatrix} \begin{pmatrix} \mathbb{1} \\ \mathbb{1} \end{pmatrix}}_{\boldsymbol{\mathcal{I}}_{\mathrm{sh}}} = \begin{pmatrix} 0 \\ 0 \end{pmatrix} \tag{2.29}$$

with the respective vector and matrix quantities $\boldsymbol{\mathcal{U}}_0^{\mathrm{Re}} = \mathbf{1} \otimes \left[\Re\{\underline{u}_0^{\mathrm{a}}\}, \Re\{\underline{u}_0^{\mathrm{b}}\}, \Re\{\underline{u}_0^{\mathrm{c}}\}\right]^{\mathrm{T}}$, $\boldsymbol{\mathcal{U}}_0^{\mathrm{Im}} = \mathbf{1} \otimes \left[\Im\{\underline{u}_0^{\mathrm{a}}\}, \Im\{\underline{u}_0^{\mathrm{b}}\}, \Im\{\underline{u}_0^{\mathrm{c}}\}\right]^{\mathrm{T}}$, $\mathbb{U}_0^{\mathrm{Re}} = \mathrm{diag}\left(\boldsymbol{\mathcal{U}}_0^{\mathrm{Re}}\right)$, $\mathbb{U}_0^{\mathrm{Im}} = \mathrm{diag}\left(\boldsymbol{\mathcal{U}}_0^{\mathrm{Im}}\right)$.

The single-phase variant of equation (2.29) can be derived in the same procedure. Calligraphic variables for three-phase quantities $\boldsymbol{\mathcal{I}}$ are exchanged by bold lowercase letters $\boldsymbol{i}$ and matrices $\mathbb{R}$ are bold capital letters $\boldsymbol{R}$. Furthermore, the simplified expressions with $\boldsymbol{J}$ and $\boldsymbol{V}$ reduce even more, such that $\boldsymbol{J} = \overline{\boldsymbol{Q}}\boldsymbol{E}_0^{\mathrm{Re}}\boldsymbol{R} - \overline{\boldsymbol{P}}\boldsymbol{E}_0^{\mathrm{Re}}\boldsymbol{X}$ and $\boldsymbol{V} = \overline{\boldsymbol{Q}}\boldsymbol{E}_0^{\mathrm{Re}}\boldsymbol{X} + \overline{\boldsymbol{P}}\boldsymbol{E}_0^{\mathrm{Re}}\boldsymbol{R}$ are introduced. The resulting equation in the decomposed representation now becomes

$$
\underbrace{\begin{pmatrix} -\boldsymbol{I}^{2N\times 2N} - \overline{\boldsymbol{E}}^{\mathrm{Im}}\boldsymbol{J} + \overline{\boldsymbol{E}}^{\mathrm{Re}}\boldsymbol{V} + \boldsymbol{G}_{\mathrm{sh}}\boldsymbol{R} - \boldsymbol{B}_{\mathrm{sh}}\boldsymbol{X} & -\overline{\boldsymbol{E}}^{\mathrm{Im}}\boldsymbol{V} - \overline{\boldsymbol{E}}^{\mathrm{Re}}\boldsymbol{J} - \boldsymbol{G}_{\mathrm{sh}}\boldsymbol{X} - \boldsymbol{B}_{\mathrm{sh}}\boldsymbol{R} \\ -\overline{\boldsymbol{E}}^{\mathrm{Im}}\boldsymbol{V} - \overline{\boldsymbol{E}}^{\mathrm{Re}}\boldsymbol{J} + \boldsymbol{B}_{\mathrm{sh}}\boldsymbol{R} + \boldsymbol{G}_{\mathrm{sh}}\boldsymbol{X} & -\boldsymbol{I}^{2N\times 2N} + \overline{\boldsymbol{E}}^{\mathrm{Im}}\boldsymbol{J} - \overline{\boldsymbol{E}}^{\mathrm{Re}}\boldsymbol{V} + \boldsymbol{G}_{\mathrm{sh}}\boldsymbol{R} - \boldsymbol{B}_{\mathrm{sh}}\boldsymbol{X} \end{pmatrix}}_{\boldsymbol{A}_{\mathrm{PF}}} \underbrace{\begin{pmatrix} \boldsymbol{i}^{\mathrm{Re}} \\ \boldsymbol{i}^{\mathrm{Im}} \end{pmatrix}}_{\boldsymbol{i}}
$$
$$
+ \underbrace{\begin{pmatrix} \boldsymbol{E}_0^{\mathrm{Re}} & \boldsymbol{0} \\ \boldsymbol{0} & -\boldsymbol{E}_0^{\mathrm{Re}} \end{pmatrix}}_{\boldsymbol{B}_{\mathrm{PF}}} \underbrace{\begin{pmatrix} \boldsymbol{p} \\ \boldsymbol{q} \end{pmatrix}}_{\boldsymbol{s}} - \underbrace{\begin{pmatrix} \boldsymbol{U}_0^{\mathrm{Re}} & \boldsymbol{0} \\ \boldsymbol{0} & \boldsymbol{U}_0^{\mathrm{Re}} \end{pmatrix} \begin{pmatrix} \boldsymbol{G}_{\mathrm{sh}} & -\boldsymbol{B}_{\mathrm{sh}} \\ \boldsymbol{B}_{\mathrm{sh}} & \boldsymbol{G}_{\mathrm{sh}} \end{pmatrix} \begin{pmatrix} \boldsymbol{1} \\ \boldsymbol{1} \end{pmatrix}}_{\boldsymbol{i}_{\mathrm{sh}}} = \begin{pmatrix} \boldsymbol{0} \\ \boldsymbol{0} \end{pmatrix} \tag{2.30}
$$

The voltage equation (2.20) in decomposed form for the single- and three-phase case becomes

$$
\underbrace{\begin{pmatrix} \boldsymbol{\mathcal{U}}^{\mathrm{Re}} \\ \boldsymbol{\mathcal{U}}^{\mathrm{Im}} \end{pmatrix}}_{\boldsymbol{\mathcal{U}}} = \underbrace{\begin{pmatrix} \mathbb{R} & -\mathbb{X} \\ \mathbb{X} & \mathbb{R} \end{pmatrix}}_{\mathbb{C}_{\mathrm{PF}}} \underbrace{\begin{pmatrix} \boldsymbol{\mathcal{I}}^{\mathrm{Re}} \\ \boldsymbol{\mathcal{I}}^{\mathrm{Im}} \end{pmatrix}}_{\boldsymbol{\mathcal{I}}} + \underbrace{\begin{pmatrix} \boldsymbol{\mathcal{U}}_0^{\mathrm{Re}} \\ \boldsymbol{\mathcal{U}}_0^{\mathrm{Im}} \end{pmatrix}}_{\boldsymbol{\mathcal{U}}_0} \quad \text{three-phase}
$$
$$
\underbrace{\begin{pmatrix} \boldsymbol{u}^{\mathrm{Re}} \\ \boldsymbol{u}^{\mathrm{Im}} \end{pmatrix}}_{\boldsymbol{u}} = \underbrace{\begin{pmatrix} \boldsymbol{R} & -\boldsymbol{X} \\ \boldsymbol{X} & \boldsymbol{R} \end{pmatrix}}_{\boldsymbol{C}_{\mathrm{PF}}} \underbrace{\begin{pmatrix} \boldsymbol{i}^{\mathrm{Re}} \\ \boldsymbol{i}^{\mathrm{Im}} \end{pmatrix}}_{\boldsymbol{i}} + \underbrace{\begin{pmatrix} \boldsymbol{u}_0^{\mathrm{Re}} \\ \boldsymbol{u}_0^{\mathrm{Im}} \end{pmatrix}}_{\boldsymbol{u}_0} \quad \text{single-phase.} \tag{2.31}
$$

Sequence-frame based approximated power flow decomposition

The sequence components for a constant power flow solution consist of three balanced decoupled systems which exactly represent its unbalanced counterpart. By a real valued decomposition and inclusion of the transformation matrix $\underline{\boldsymbol{T}}_{\mathrm{pnz}}$ the approximation (2.29) can be extended. Because both the decomposed approximated power flow (2.29) and the transformation (2.28) are linear, the sequence-frame voltage $\boldsymbol{\mathcal{U}}_{\mathrm{pnz}}$ of every node in the grid can be expressed as a linear function of the decomposed apparent power $\boldsymbol{\mathcal{S}}$ as

$$
\boldsymbol{\mathcal{U}}_{\mathrm{pnz}} = \underbrace{\begin{pmatrix} \Re\{\underline{\mathbb{T}}_{\mathrm{pnz}}\} & -\Im\{\underline{\mathbb{T}}_{\mathrm{pnz}}\} \\ \Im\{\underline{\mathbb{T}}_{\mathrm{pnz}}\} & \Re\{\underline{\mathbb{T}}_{\mathrm{pnz}}\} \end{pmatrix}}_{\mathbb{K}_{\mathrm{PF}}} \underbrace{\left(\mathbb{C}_{\mathrm{PF}}\left(\mathbb{A}_{\mathrm{PF}}^{-1}\mathbb{B}_{\mathrm{PF}}\boldsymbol{\mathcal{S}} - \mathbb{A}_{\mathrm{PF}}^{-1}\boldsymbol{\mathcal{I}}_{\mathrm{sh}}\right) + \boldsymbol{\mathcal{U}}_0\right)}_{\underbrace{\begin{pmatrix} \boldsymbol{\mathcal{U}}^{\mathrm{Re}} \\ \boldsymbol{\mathcal{U}}^{\mathrm{Im}} \end{pmatrix}}_{\boldsymbol{\mathcal{U}}}} \tag{2.32}
$$

with $\underline{\mathbb{T}}_{\mathrm{pnz}} = \mathrm{diag}\left(\underline{\boldsymbol{T}}_{\mathrm{pnz}}, \ldots, \underline{\boldsymbol{T}}_{\mathrm{pnz}}\right)$. Note that $\boldsymbol{\mathcal{U}}$ is a solution of (2.29) which means that (2.32) gives an explicit relationship between sequence-components in current and voltage and the apparent power of each node.

Furthermore, it is assumed that the slack bus voltage is balanced and thus does not contribute to the negative- or zero-sequence component. Consequentially, only the voltage drops across the branches of the grid and the neutral conductor must be considered. These voltage drops in sequence-frame can be expressed as

$$\Delta\boldsymbol{\mathcal{U}}_{\text{pnz}} = \mathbb{K}_{\text{PF}} \underbrace{\left(\mathbb{C}_{\text{PF}} \left(\mathbb{A}_{\text{PF}}^{-1}\mathbb{B}_{\text{PF}}\boldsymbol{\mathcal{S}} - \mathbb{A}_{\text{PF}}^{-1}\boldsymbol{\mathcal{I}}_{\text{sh}}\right)\right)}_{\Delta\boldsymbol{\mathcal{U}}} \tag{2.33}$$

where $\Delta\boldsymbol{\mathcal{U}}_{\text{pnz}}$ is the decomposed form of the voltage drops in sequence-frame for every node. Moreover, the individual sequence-voltage components extracted from (2.33) can be expressed by the linear mapping (2.34)

$$\begin{aligned} \Delta\boldsymbol{\mathcal{U}}_{\text{pos}} &= \mathbb{M}_{\text{pos}}\Delta\boldsymbol{\mathcal{U}}_{\text{pnz}} \,\text{with}\, \mathbb{M}_{\text{pos}} = \text{diag}\left(\mathbf{1}\right) \otimes \left[1\,0\,0\right]^{\text{T}} \\ \Delta\boldsymbol{\mathcal{U}}_{\text{neg}} &= \mathbb{M}_{\text{neg}}\Delta\boldsymbol{\mathcal{U}}_{\text{pnz}} \,\text{with}\, \mathbb{M}_{\text{neg}} = \text{diag}\left(\mathbf{1}\right) \otimes \left[0\,1\,0\right]^{\text{T}} \\ \Delta\boldsymbol{\mathcal{U}}_{\text{zero}} &= \mathbb{M}_{\text{zero}}\Delta\boldsymbol{\mathcal{U}}_{\text{pnz}} \,\text{with}\, \mathbb{M}_{\text{zero}} = \text{diag}\left(\mathbf{1}\right) \otimes \left[0\,0\,1\right]^{\text{T}} \end{aligned} \tag{2.34}$$

Note that (2.33) expresses the sensitivity of the sequence-components $\Delta\boldsymbol{\mathcal{U}}_{\text{pnz}}$ of each node by the injection of apparent power.

Operating-point accuracy and absolute error evaluation

To compare the discussed methods a simple two node example is constructed with one cable as depicted in Figure 2.2. The used cable is of type NAYY $4 \times 70\,\text{mm}^2$ with a length of 750 m. Three operating points which have been determined with the non-linear FBS method are used to compare the approximation methods. It should be noted that only methods which use the voltage magnitude are compared, because the others are using PMUs.

Table 2.2: Calculated operating points of the two node example using the non-linear FBS method.

Scenario	**Active power**	**Reactive power**	$\lVert\underline{u}_{\text{FBS}}\rVert$	$\measuredangle\underline{u}_{\text{FBS}}$
1. High loading	-0.4 p.u.	0.2 p.u.	0.86 p.u.	-5.44
2. High in-feed	0.6 p.u.	-0.2 p.u.	1.14 p.u.	4.92
3. High reactive power	0.0 p.u.	-0.4 p.u.	0.96 p.u.	7.47

The nodal power is varied linearly for both active and reactive power in the range of {-0.6 p.u.,...,0.6 p.u.} with respect to the operation point specified in Table 2.2. The power base is set to 50 kVA and the voltage base to $400/\sqrt{3}$ V. The power variation is plotted relative to the horizontal axis as Δp, Δq while 0 represents the apparent power that is equal to the operating point.

The error resulting from the approximations is evaluated based on

$$\Delta \left\|\underline{u}_{\text{Error}}\right\| = 100\% \frac{\left\|\underline{u}_{\text{FBS}}\right\| - \left\|\underline{u}_{\text{Approx}}\right\|}{\left\|\underline{u}_{\text{FBS}}\right\|}, \quad \measuredangle \underline{u}_{\text{Error}} = 100\% \frac{\measuredangle \underline{u}_{\text{FBS}} - \measuredangle \underline{u}_{\text{Approx}}}{\measuredangle \underline{u}_{\text{FBS}}}$$

with $\underline{u}_{\text{Approx}}$ the voltage values that result from the approximations. In this way, the absolute accuracy of the power flow approximation and the sensitivity with respect to the operating point can be investigated. The vertical axis plots the error in percent with respect to the non-linear solution.

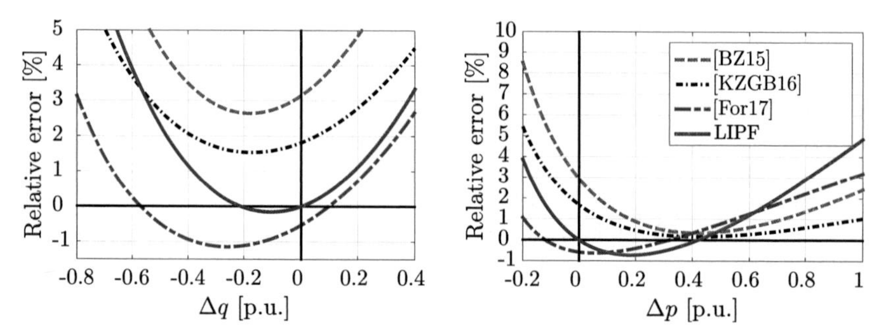

Figure 2.11: Voltage magnitude error of the different power flow approximations with respect to an operating point, scenario 1

Table 2.3: Error results at the operating-point for scenario 1: high load

Reference	Line color	Mag. accuracy	Angle accuracy
[KZGB16]	black dot-dashed	3.09 %	N.A.
[BZ15]	magenta dashed	1.78 %	−3.0 %
[For17]	red dash-dash	−0.57 %	N.A.
This Work	blue solid	0.08 %	3.44 %

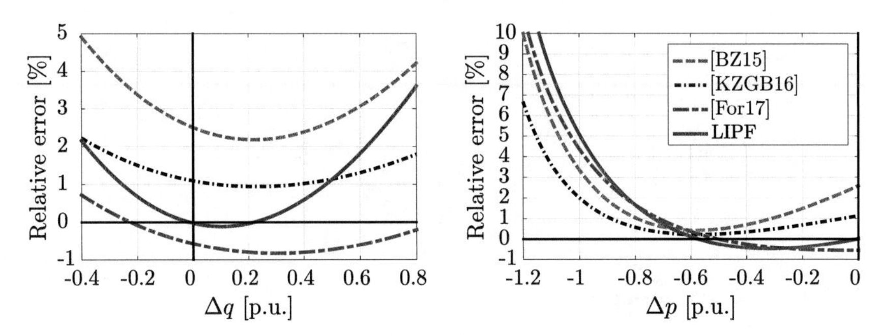

Figure 2.12: Voltage magnitude error of the different power flow approximations with respect to an operating point, scenario 2

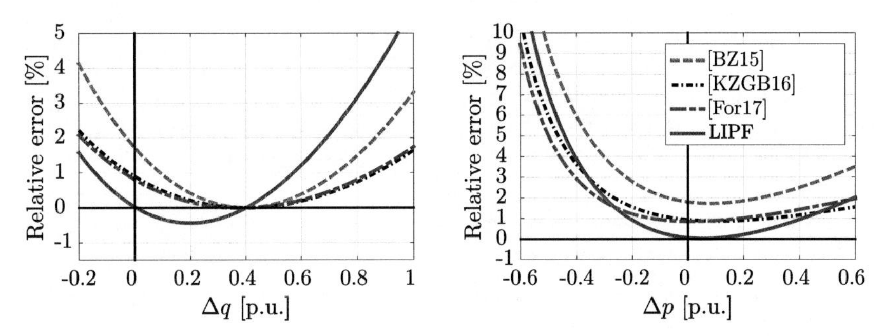

Figure 2.13: Voltage magnitude error of the different power flow approximations with respect to an operating point, scenario 3

Table 2.4: Error results at the operating-point for scenario 2: high in-feed

Reference	Line color	Mag. accuracy	Angle accuracy
[KZGB16]	black dot-dashed	2.54 %	N.A.
[BZ15]	magenta dashed	1.11 %	−2.48 %
[For17]	red dash-dash	−0.54 %	N.A.
This work	blue solid	$1.4e^{-3}$ %	3.77 %

Table 2.5: Error results at the operating-point for scenario 3: high reactive power injection

Reference	Line color	Mag. accuracy	Angle accuracy
[KZGB16]	black dot-dashed	1.81 %	N.A.
[BZ15]	magenta dashed	0.95 %	−1.8 %
[For17]	red dash-dash	0.85 %	N.A.
This Work	blue solid	0.07 %	1.95 %

Intermediate conclusion

The numerical results depicted in Figure 2.12-2.13 show that the LIPF method outperforms the other methods in terms of voltage magnitude accuracy. Especially in operating points where reactive power is injected or drawn from the grid the accuracy is much higher.
In the case of phase angle accuracy, the LinPF [BZ15] performs almost equally, which indicates that this method is very useful for phase angle estimation. This can be explained with the low X/R ratio found in low voltage grid. The phase angle is coupled with the voltage drop across the inductance of the line. So neglecting the losses, as has been done with the LinPF [BZ15] does not have a large influence on the calculated angle. In summary, the following result has been achieved:

- An approximated power flow method was presented that can use voltage magnitude and power measurements to increase the accuracy of the power flow result.
- It is modular as it can be parameterized off-line or on-line depending on available communication, measurements and accuracy requirements.
- Especially for high loading or large reactive power in-feed the method, performs better than current literature.

It should be noted that the increased accuracy of the LIPF stems from the fact that an approximate solution of the power flow from (2.29) and (2.30) is solved to generate the matrices. This approximate solution is realized with one matrix inversion. The accuracy of the method can be further increased if multiple of these matrix inversions are calculated in a sequence.

3 Power system grid setup for simulation and practical demonstration

This chapter introduces the power system benchmarks which will later be used to investigate the behavior and performance of the control and estimation algorithms. They are separated into simulation and practical demonstration and will be introduced briefly. Parameters of the power lines and substation transformers are presented in the appendix B. This sequence is chosen to first establish a common base with the benchmarks and introduce different cases for the simulations later.

Moreover, for the verification of the results two different power flow solvers have been used in this thesis. The first is the free software package called OpenDSS [MHR12], which is widely used throughout academia as a reference solver. Especially for the three-phase investigations this platform has been used to verify the results. The second power flow solver is a Forward-Backward-Sweep method [Zim95] developed in MATLAB© which was introduced in Section 2.3.2.

3.1 Simulation benchmarks for validation

The following two sections introduce the low voltage distribution grids that build the foundation for the simulation based validation. Every grid is used to validate a specific property of the algorithms.

3.1.1 Four node minimal example grid

To demonstrate the behavior of the balanced operation algorithm presented in Section 4, a very simple example grid with only four nodes is introduced as depicted in Figure 3.1. It is used to demonstrate the constraint satisfaction of the proposed DMPC operation and the battery storage usage in an extreme scenario. The branches of the example grid are

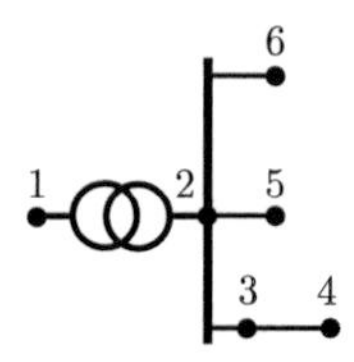

Figure 3.1: Simple four node low voltage benchmark grid

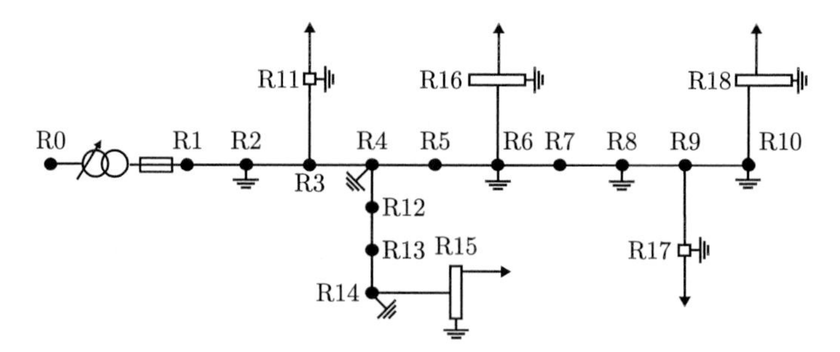

Figure 3.2: CIGRÉ residential low voltage grid (figure adapted for this thesis)

represented with standard cable types N(A)YY with $95\,\text{mm}^2 - 180\,\text{mm}^2$. The grid does not fully represent a realistic setup but is rather a minimal example to demonstrate the overall behavior of the algorithm. Cable types and branch length as well as data related to the substation transformer, can be found in Appendix B, Table B.1 and Table B.2.

3.1.2 CIGRÉ residential low voltage grid

The CIGRÉ task-force C6.04.02 developed several medium and low voltage grids as benchmarks for time series based power flow simulations [Cig14]. Especially for low voltage grids there are three benchmarks for residential, industrial and commercial grid structures. In this work only the residential part of the grid benchmarks as depicted in Figure 3.2 is used for further investigations as it is the closest to represent a rural area in Europe. It should be mentioned that for a German grid the cable core section of this benchmark is comparable to the standard cable type N(A)YY with $120\,\text{mm}^2 - 180\,\text{mm}^2$ [Lan13]. The CIGRÉ grid is thus used to study the effect of battery storage and electric vehicles on the voltage and current control mechanisms. The example setup which was described in [PHS05] is used in this thesis for the investigation. Parameters of the benchmark can be found in Appendix B,Table B.1 and Table B.3.

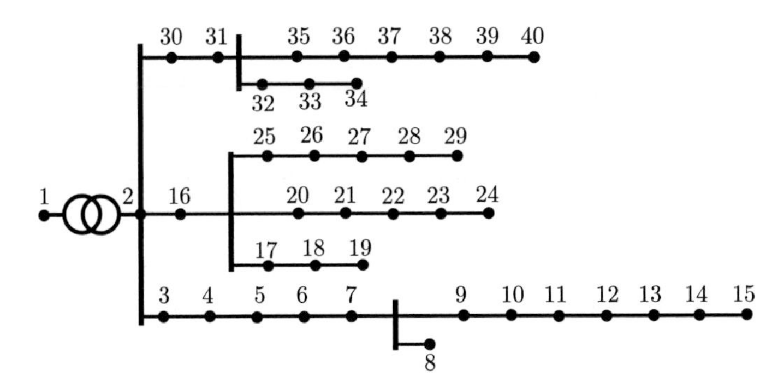

Figure 3.3: Rural low voltage grid derived from a realistic LV distribution grid

3.1.3 German rural low voltage grid

A second realistic example of a low voltage grid is constructed from the planing data of a real rural power distribution grid. The investigated grid has 40 nodes and three main spurs originating from the substation transformer as depicted in Figure 3.3. This example is used to investigate problems related to rural grids such as undetectable over currents in the branches and over-voltages at the end of long feeders. For this purpose it has one long feeder with a dairy farm at node eight were usually high photo-voltaic penetration can be assumed. Additionally, several shorter cable sections between nodes 9-15, where electric vehicle can be connected and as such possible over-currents can be studied. Furthermore, the grid will be used for the unbalance mitigation strategy.

The upper two branches of the grid have homogeneous cable lengths. Again the parameters of the benchmark can be found in Appendix B, Table B.1 and Table B.5.

3.2 Smart-grid laboratory for verification

To evaluate the feasibility of the proposed balanced control strategy in a realistic setup, a hardware grid available at the Institute for Control Systems was fitted with additional hard- and software components to work as a modular smart-grid as depicted in Figure 3.4. Not only is it possible to conduct experiments with a scaled medium voltage grid, but additionally an emulator was implemented on a rapid prototyping system (RPS) to account for arbitrary voltage levels and grid structures. The RPS is a dSpace©

product which can be configured with different interfaces and processors.

The interface between the dSpace© system and the measurements, inverter and gen-

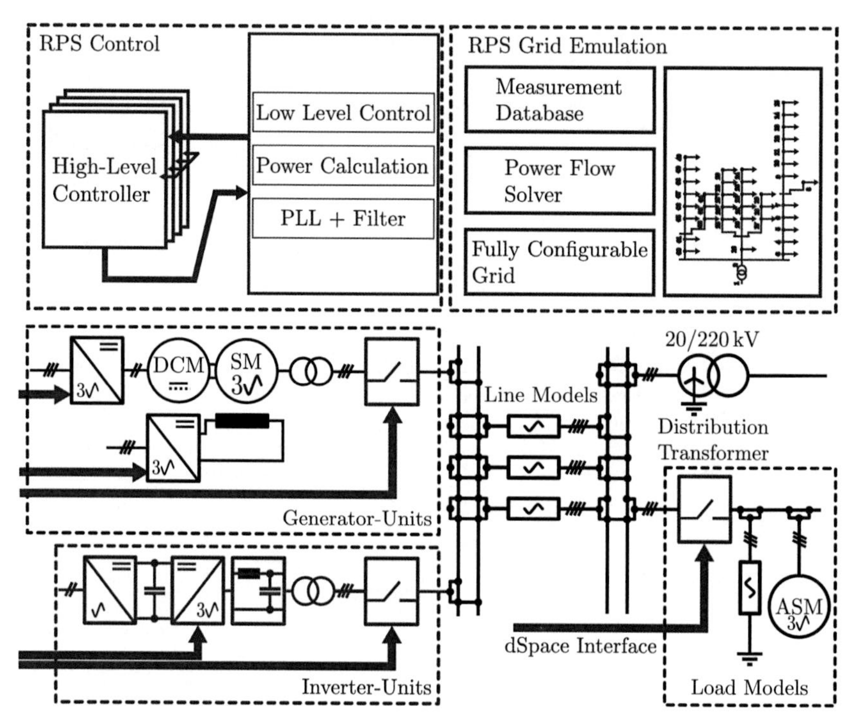

Figure 3.4: Block diagrams of the modular smart-grid demonstrator with controller and emulation platform at the top and fitted hardware grid at the bottom

erator control panel was developed for this work. To achieve this, the analog digital converters (ADC) and digital analog converters (DAC) were split from their standard dSpace© connector to several Sub-D 9 connectors. An interface card was designed that can be used for different purposes like control and measurement. From this Sub-D 9 each of the components depicted in Figure 3.4 can be connected via a designed measurement cable. The grid switches on the other hand are connected to the digital I/O cards of the dSpace© system. Again a special cable was constructed for this purpose.

Elements in Figure 3.4 which have been developed in this thesis are depicted in blue color. Hardware parts which were available in the grid demonstrator are depicted in black color.

3.2.1 Medium voltage hardware demonstrator

The heart of the demonstrator is a customized three-phase four-wire scaled hardware grid for the evaluation of 220 kV overhead transmission and 20 kV cable based distribution grids.
For the investigation in this thesis the grid is operated as a 20 kV cable based distribution grid with direct grounded neutral connection at the transformer. This neutral connection type is chosen to suppress unbalance in the nodal voltages which result from parameter variation in the simulated cables. The cables are available in two lengths with 4 km and 8 km and are realized as concentrated components in a Π equivalent circuit. Additionally, the grid model has several three-phase impedances as capacitor, inductors and resistors which can be configured with different settings and an induction machine to simulate loads.
The scaling between real world and model values is summarized in Table 3.1.

Table 3.1: Grid demonstrator scaling parameters if used as a 20 kV cable grid

Component	Model value	Real value
Voltage	20 kV	220 V
Current	422 A	4.3 A
Apparent Power	14.5 MVA	1600 kVA
Impedance	27.6 Ω	30.3 Ω

Synchronous machine based generation units

The smart-grid is equipped with two 12 kVA synchronous machines (SYM) which can be used to emulate arbitrary generation units and flexible demand. Each SYM is on the same shaft as an inverter driven direct current machine (DCM) that provides the torque for active power generation and works as a controllable load if active power is drawn from the grid. The low level controller function for excitation of the SYM and speed control of the DCM are embedded in their respective inverters and can either be set to automatic or be directly controlled by the dSpace©. Furthermore, both units are coupled to the grid via an electromechanical switch in order to automatically synchronize and operate them (see Figure 3.4 middle left).

Inverter based generation units

To account for large battery storages or photo-voltaic plants as well as inverter connected wind-power plants in medium voltage distribution grids an $2\,\mathrm{kVA} \equiv 18.125\,\mathrm{MVA}$ two-level inverter from ST-Microelectronics (STEVAL-IHM028V2) [Mic17] was equipped with a micro-controller (STM32F4DISCOVERY) [STM17] and an isolated analog interface for the dSpace© interconnection (see Figure 3.4 lower left). The μ-controller receives the commanded output voltages as u_α and u_β values – see [Sch15a] for a review on control methods and transformations – from the dSpace© low level control. For the grid connection an LC-filter circuit is used that produces a nearly sinusoidal output voltage.

Three-phase current and voltage measurement units

For the different low- and high level controller layers a suitable measurement signal of voltage, current and power must be supplied to the dSpace©. A customized three-phase measurement device was developed to account for this need, taking into account the bandwidth needed for the inverter based units. To assure a high bandwidth and good overload feature magneto-resistive sensors were implemented for the current sensing.

3.2.2 Rapid prototyping system

For the control verification and the power grid emulation a dSpace© DS1006© rapid-prototyping system is used. The following section describes the two applications that run on the processing units.

Multi-core distributed controller platform

All low and high level control structures as depicted in Figure 3.4 are implemented in a Simulink© environment as embedded MATLAB© functions. For the high level controller implementation the individual four physical cores of the DS1006© are setup as a multiprocessor system and communicate over a common bus which simulates the communication interface. This interface would be power line communication or some form of wireless technology in a real grid. For the remainder of this thesis, the communication infrastructure is not simulated.

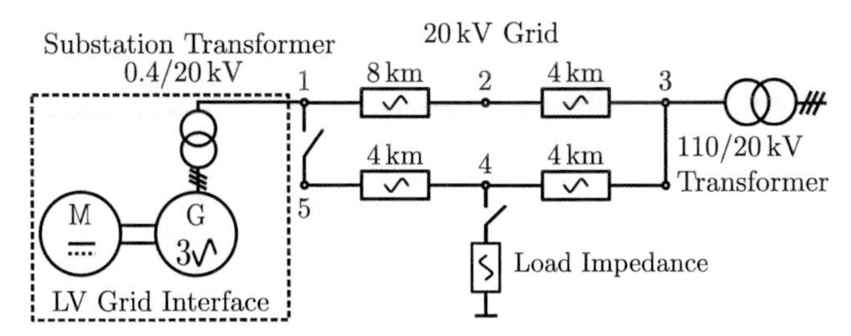

Figure 3.5: Experimental setup of the medium voltage grid with one of the machine units acting as a simulated substation transformer, interfacing emulated low voltage grid and medium voltage hardware demonstrator ©IEEE2018

Low voltage grid hardware emulation

Because the hardware demonstrator can only be used to simulate high- or medium voltage grids, a second dSpace© was used to emulate a low voltage grid by implementing a real-time grid simulator. This is achieved by solving the power flow problem of the grid repetitively with a three-phase four-wire FBS method that retrieves power inputs from an interface card or a serial communication. Uncontrolled nodes on the other hand generate their load profile from an internal database. Furthermore, the interface between the controller RPS and the emulation RPS is established by analog signals.

3.2.3 Medium voltage grid setup for the controller verification

To investigate the effect of monitoring the interconnection between MV and LV grid a cable based 20 kV benchmark grid is constructed as shown in Figure 3.5. The substation transformer of the LV grid is simulated with one of the SYM/DCM machine units that can act as generator and motor alike. At this point the interface between the emulated LV grid in the RPS and hardware demonstrator is established.

The nodal voltages and currents of the MV grid are directly acquired by measurement units that transform the real world values to a level that can be directly processed by the RPS.

3.3 Prediction of photo-voltaic in-feed and generation of load profiles

To sample real load and in-feed profiles a smart-meter was developed based on the SOCOMEC A10 network analyzer and a Raspberry Pi model 1B. This smart-meter is able to measure three voltages and three currents as well as the sum of active and reactive power. All measurements are sampled with a granularity of one second. The data base of the individual devices was used to acquire different load profiles of low voltage consumers in the field.

A measurement campaign was conducted in a dairy farm and a hotel to classify the load profile and setup prediction models for the operation. The results showed that especially mean values of the dairy farm over several weeks can provide a simplified mean value prediction. For the dairy farm it was found that seasonal changes were insignificant. In case of the hotel these changes could be observed, but where much less severe than expected. This could be explained by the small number of rooms and rather large restaurant and seminar space. The restaurant was occupied very frequently which lead to a high activity in the kitchen. Cooking and preparation was thus identified as a major load factor in the hotel.

Furthermore, the measurement data from the two photo-voltaic plants that were investigated, showed that heir power in-feed into the grid could be approximated by using a sinusoidal fitting on the measurement data as in [HEAF10]. Because the principal clear sky irradiation is only changing slowly over the year [WS09], the validity window for the approximation is one to several weeks.

3.3.1 Demand and in-feed data in the rapid prototyping system

The demand data from the IEEE European low voltage feeder consists of 100 different load profiles for domestic consumers with a granularity of 1 min. These profiles are phase specific and can thus be combined arbitrary for the different phases, as depicted in Figure 3.6 c). In case of balanced single-phase investigations, the positive sequence component is calculated from theses load profiles for each scenario. A few of these load curves are depicted in Figure 3.6. Aside from residential household consumers, rural areas can be quite diverse. To account for this diversity, hotels and dairy farms are integrated as an additional load form which is usually present in the grid as well. Depicted in Figure 3.6 are the load curves of a generic hotel and a dairy farm, in a) in red and blue line color. The scaled measurements of injected power from photo-voltaic generation are available for different days and months and each purpose. They can be

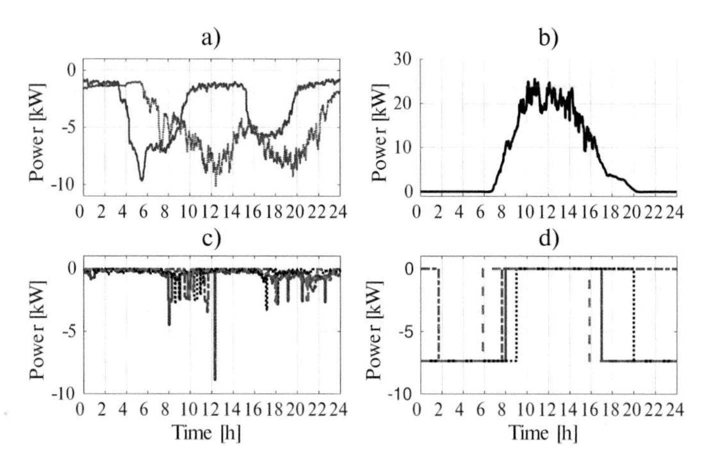

Figure 3.6: Exemplary unscaled single-phase power demand curves for a hotel, dairy farm, domestic demand and generic EV charging cycles ©IEEE2018

connected in three- or single-phase depending on the use case. An exemplary profile is depicted in Figure 3.6 b). Via generic charging cycles, electric vehicles are introduced. These charging cycles have been generated for 3.6 kVA and 11 – 22 kVA chargers and represent the maximum charging capability of the electrical inverter in the cars. Note that electric vehicle chargers for 3.6 kVA are assumed to be connected in single-phase while 11 – 22 kVA chargers are connected in all three phases. All data sets for the charging profiles are available in a time granularity of one minute. Generated profiles for the generic charging cycles are shown in Figure 3.6 d)

The measurements for the data sets are saved in the flash of the dSpace© system. This way, the load, and distributed generation can be emulated in real-time. Due to the limited storage capacity of the RPS, the measurements can only be stored with one-minute granularity. If smaller sample times of the RPS are needed, the measurements have to be up-sampled to fit the respective simulation cycle of the model. In between two respective measurements with a distance of one minute, a linear interpolation was used. Additionally, white noise in the range of 5 % of the last value was added to the linearly interpolated values.

4 Dispatch of balanced generation units and flexible demand in low voltage grids

This section introduces a framework that combines dynamic optimal power with distributed model predictive control. This framework is used to regulate the dispatch of distributed generation and demand response of electric vehicles for grid operation purposes. The focus of the operation strategy is a maximization of usable distribution capacity of the grid, while always maintaining system limitations. The ability of distributed model predictive control to operate closely to system boundaries is exploited to increase the grid transportation capacity. An efficient implementation whose feasibility is shown in an experimental setup is achieved by eliminating the power flow constraints with the developed power flow approximation. This section is based on the publications [BFB⁺16, BBL18].

4.1 Introduction and related work

The operational possibilities that are introduces with renewable generation, stationary battery storages and flexible demand like electric vehicles are very promising. At the same time the digital transformation of power grids with an additional communication infrastructure and sensing equipment paves the way for grid estimation, new control approaches for nodal voltage and reactive power support in distribution grids. In the literature three groups of approaches for the operation of distribution grids can be identified:

- Conventional centralized, with one control room at the network operator side. At a central point, all measurements are collected, and the actuation signals originate from this place as well. [Oer14, Alz17, For17].
- Decentralized control or operation, where only local inverters connect to the grid. In some cases, there are combinations of several components, like battery storage, electric vehicle, or others. However, there is no information exchange between the controllers [Bra06, AP13, Ste13].

- Methods based on distributed control with several controllers interacting with each other. Every controller operates a single node or a certain part of the grid [Neg07, AYVM+09, BZ11, AZG13].

Especially distributed approaches combine the advantages of both decentralized and central control and at the same time, mitigate the disadvantages. While decentralized control can be unstable due to missing knowledge about the grid, no communication, or controller saturation [ABK+15, CBCZ16], central control systems are inflexible and have a single point of failure behavior. In power systems, distributed control can increase the security of supply, because it has inherent robustness to failure of single controllers. Furthermore, measurements do not have to be transported way back to a central point, but can instead be preprocessed or aggregated. The aggregation reduces the loading of both the back-end and the communication infrastructure.

A great deal of research was spent on distributed optimization methods that use a model of the power grid in the form of a power flow formulation, see [MDS+17] for an in-depth review. The network model makes it possible to determine the grid behavior in advance and to estimate the effect of individual participants on an exact basis. Especially, the alternate direction method of multipliers (ADMM) and it is successors play a vital role in the field of grid operation with both power flow approximations [BCCZ15, GDC+16, MWGC16] and for the nonlinear non-convex formulation [Ers14]. However, the communication demand, in order to converge to a solution, is generally very high. Furthermore, it is challenging to design approaches, where the optimization problem is feasible after each distributed solution. Thus, in case of a communication failure, no plant-wide solution that satisfies the constraints of the power system would be available. Furthermore, the presented methods have to be classified as open loop, as no information about the power system is integrated on-line.

Another exciting research direction that has emerged in recent years is the closed-loop real-time distributed optimal power flow [GL16, HBH16, ANPS+16]. The main idea of the approaches is to further reduce the need for a model by using the power system itself as a real-time solver. In an iterative procedure, the reaction of the grid to a control signal is measured. The measurements are used as feedback for the calculation in the next iteration. The distributed solution is achieved by first-order gradient methods [Tsi13] and their extension, which are computationally easy to solve. Their main drawback, however, is the lack of inclusion for future time instances, e.g., for a dispatch prediction. However, this feature is currently seen as a necessity by system operators for efficient planning [Roa16].

Until now, the discussed methods were not dedicated to voltage control but rather designed for operation management. Reactive power support and voltage control are closely related, but in order to regulate the voltage, sometimes controller decisions counteract economic goals which are embedded in the objective of the optimal power flow. Some authors formulate the voltage control as a reference tracking problem in a distributed fashion [KZGB16]. However, tracking a reference is counterproductive if the operational capabilities of the grid should be maximized. Instead, set-points for reactive power support are favored. Sometimes this requires operating close to the boundaries. A semidefinite programming approach (SDP) was presented in [ZLDGT14] that can handle this operation strategy, but economic goals cannot be considered anymore. Especially when the controller task is to keep the voltage in predefined boundaries, formulated by constraints of the optimization problem. To the authors best knowledge, the use of distributed model predictive control, to combine several features of voltage control and operation management has only been proposed in very few publications [Neg07, Mor13]. Both monographs only investigate the general possibilities of distributed model predictive control on power networks but did not explicitly exploit its advantages for optimal power flow. Furthermore, the explicit use of power flow approximations is not investigated.

4.2 Distributed control setup

The distributed control structure for the operation management and reactive power support developed in this work consists of several local controllers, that either control a small part of the grid, or a single node as depicted in Figure 4.1. Together they are embedded in a distributed model predictive control framework and jointly solve a feasible cooperation based optimal dispatch problem [VRW06, Ven06]. Information between the controllers is transmitted via a common bus, that allows for a full-duplex exchange of data. Moreover, the local controllers have full access to inverters of the distributed generators, battery storages, and electric vehicles, if they are connected to the grid in all three phases. For the feasible cooperation, the local knowledge of each controller must be sufficient such that he can issue a solution to the optimal power flow problem, which is valid for the whole operated power grid. This local knowledge consists of a power flow model, which reflects the stationary grid behavior. Usually, for a full model of the power grid, a three-phase power flow must be formulated and solved inside the optimization problem. However, for a real-time implementation, this is not a feasible direction, which can be observed from the complexity of methods that implement the full three-phase AC model [BLR$^+$11, APCP13]. So instead of the full representation, the grid is modeled as an approximated balanced single-phase equivalent model (2.30) that reflects only the larger units connected to the grid. This representation is sufficient because according to [VDE18], new PV generation units with a rated power 4.6 kVA$\leq$, which have the main influence on the overall grid behavior are included. The mean value of apparent power for all three phases is used to approximate the balanced loading.

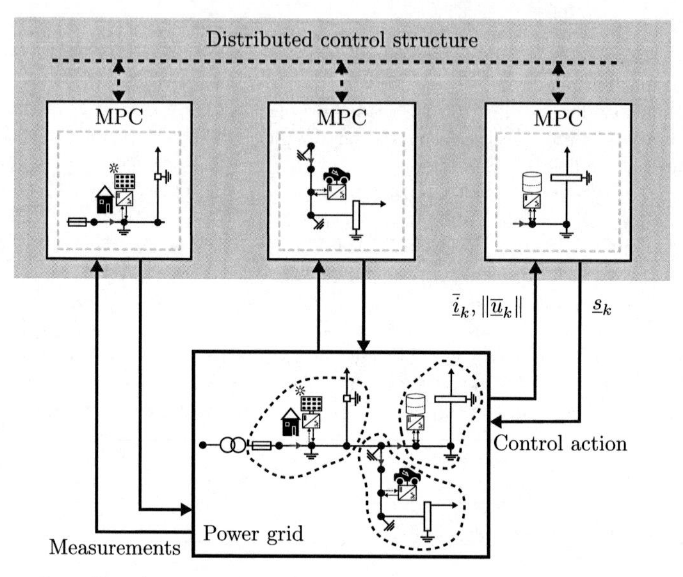

Figure 4.1: Distributed model predictive control structure for the balanced operation of the low voltage grid

The distributed controller environment provides privacy for every customer when it comes to information exchange between them. This privacy is achieved with the construction of performance charts, that reflect the load, current PV in-feed power, and available flexibility form battery storages. These performance charts are usually used to classify generators based on electric machines but can be very use-full in describing the capabilities of the local grid segments. Furthermore, as discussed before, they reflect the possible contribution and ancillary services that each controlled segment can provide for the grid, but do not compromise the exact values of its components. For the integration in an MPC scheme, single values are not sufficient and instead predictions of load and generation must be provided locally in order to generate the performance charts for several steps up to the prediction horizon. The mechanism to generate these profiles have been presented in Section 3.3 and are only referred here.

Limited knowledge results in a trade-off between maximizing local needs of each consumer and the common global interest of a grid operation without high transmission losses. Each controller only has a local model of his components and consequently has to approximate the local objective of others. The approximation is achieved, by assuming that the goal of other controllers is to in-feed power from distributed generation, to receive the current compensation and secondly, to reduce energy costs by minimizing its demand. Components like wear costs of batteries are thus not included in the global point of view but considered in a local objective. Through this splitting of the objective function for each controller into a global and local part, the controllers have a common understanding of the grid operation perspective, but still, have the freedom to push a solution that only has a benefit for them self. Using these two objectives, the controllers solve the dynamic optimal power flow in a distributed fashion in a four-step negotiation approach. With the four steps, they agree on a proposal that has the highest benefit for the grid and the lowest additional cost for their local interests. In each step, there is only one broadcast of information to every other sub-system, which keeps the communication to a minimum.

In summary, the distributed method combines different ancillary services for the operation of the grid. Following main novel features are contributed:

- Privacy respecting communication between local controllers, by means of a transformation. Available flexible power (positive and negative) is combined with the demand and in-feed predictions and shared with other controllers in the form of performance charts. With the help of these performance charts the local controllers decide which generation unit should provide ancillary services or needs to reduce its in-feed due to system limitation.

- Robustness against controller failure, by a special setup of the local MPC problem. This is achieved by giving each controller global information of the grid in form of a power flow model and knowledge about the flexibility each section of the grid can offer. With this information each controller can find a feasible solution for the power flow in grid. System limitation can be kept by all means, even if one controller has to go off-line or has a unplanned failure.

- Exploiting the low voltage grids full power transmission capabilities by operating the grid close to system limitations if necessary. At the same time active power curtailment is only used as a last resort and the use of reactive power support is reduced.

- Inclusion of an explicit linear power flow formulation that can be substituted into the objective function and thus must not longer be considered in the constraints. Furthermore, the formulation gives additional accuracy at extreme situation e.g. very close to the lower voltage boundary, where the pu-curve already has a strong curvature, as depicted in Figure 2.6.

4.2.1 Local setup for each controller

It is assumed that through the available base meters, the interaction of each node with the grid is measurable. Furthermore, the respective measurements are available in a higher or at least the same time granularity as the controller cycle time. This assumption is reasonable, as smart-meter-gateways and base meters can already support tariff profiles, that can handle measurement intervals smaller than 15 minutes [BSI19]. Repetitive load profiles for the prediction of household, hotel or dairy farm are generated locally by each of the controllers, based on windowed data, until a certain time instant N_{past} in the past. The measurements $\overline{p}_{\text{L},k-l}, \overline{q}_{\text{L},k-l}$ with $l = \{N_{\text{past}}, \cdots, 1\}$ are acquired by the base meters. k represents the current step in time.

In the following these prediction are marked with a hat, as $\hat{\boldsymbol{p}}_{\text{L},k} = [\hat{p}_{\text{L},k}, \hat{p}_{\text{L},k+1}, \cdots, \hat{p}_{\text{L},k+N_{\text{P}}}]^{\text{T}}$, $\hat{\boldsymbol{q}}_{\text{L},k} = [\hat{q}_{\text{L},k}, \hat{q}_{\text{L},k+1}, \cdots, \hat{q}_{\text{L},k+N_{\text{P}}}]^{\text{T}}$ as an example for the load predictions. For

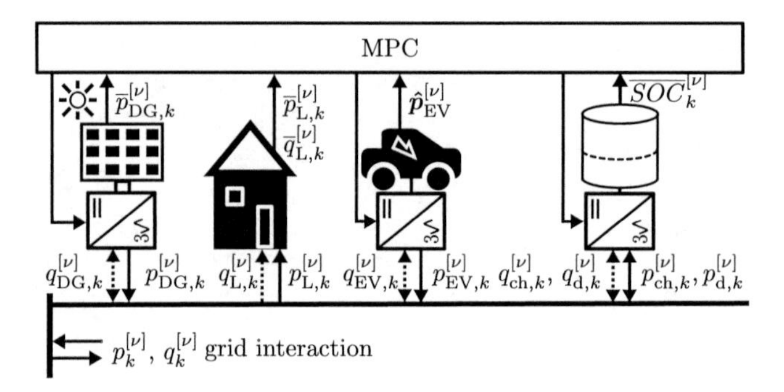

Figure 4.2: Exemplary local node setup with local MPC controller, PV generation unit, domestic demand, EV and battery storage.

the distributed generation an approach is chosen, that is based on a fitting technique which relies on inverter measurements [HEAF10] instead of using the clear sky formulation and it's extensions [WS09]. It is used to generate the predictions for the distributed generation $\hat{\boldsymbol{p}}_{\text{DG},k}$. In case of the electric vehicles, the situation is different, as it is not possible to predict the arrival of the domestic vehicle owner. Thus, the charging controller inside the car communicates with the MPC and uploads a maximum charging curve and reactive power flexibility $\hat{\boldsymbol{p}}_{\text{EV},k}, \hat{\boldsymbol{q}}_{\text{EV},k}$ as soon as the electric vehicle is connected for charging. The feedback of the battery storage to the local controller is the current measured state of charge $\overline{SOC}_k$.

Internal power distribution

To interact with the grid every local MPC in node $[\nu]$ can adjust the active and reactive output power of flexible units. **L**, **DG** and **EV** denote load, distributed generators and electric vehicles respectively. All of the flexible units are connected to the grid via inverters as depicted in Figure 4.2. The power balance equation of each node has the following form

$$p_k^{[\nu]} = p_{\mathrm{L},k}^{[\nu]} + p_{\mathrm{DG},k}^{[\nu]} + p_{\mathrm{EV},k}^{[\nu]} + p_{\mathrm{ch},k}^{[\nu]} + p_{\mathrm{d},k}^{[\nu]} \tag{4.1}$$

$$q_k^{[\nu]} = q_{\mathrm{L},k}^{[\nu]} + q_{\mathrm{DG},k}^{[\nu]} + q_{\mathrm{EV},k}^{[\nu]} + q_{\mathrm{ch},k}^{[\nu]} + q_{\mathrm{d},k}^{[\nu]} \quad \forall \nu \in \mathcal{N} \tag{4.2}$$

where $p_k^{[\nu]}$ and $q_k^{[\nu]}$ are the local active and reactive power exchange of node ν with the grid. The indexes **ch**, **d** stand for charge and discharge power of the battery storage. Both for reactive and active power two separate variables $q_{\mathrm{ch},k}^{[\nu]}, q_{\mathrm{d},k}^{[\nu]}$ are introduced even though they are provided by the same inverter. This is needed in the controller to treat the in-feed and consumption of the battery storage separately with different cost terms. Equations (4.1) and (4.2) are also called the local power split.

Battery storage model

The the battery storage at node ν is modeled as a discrete-time integrator with direct power inputs. With this formulation different mechanisms like discharging losses or aging of the battery, as well as any nonlinear cell behavior is neglected [For17]. The overall efficiency is generated by adding each of the individual efficiencies from inverter and battery. The discrete-time equation has the following form

$$SOC_{k+1}^{[\nu]} = SOC_k^{[\nu]} + \eta_{\mathrm{ch}}^{[\nu]} T_{\mathrm{s}} p_{\mathrm{c},k}^{[\nu]} + \frac{T_{\mathrm{s}}}{\eta_{\mathrm{d}}^{[\nu]}} p_{\mathrm{d},k}^{[\nu]} \tag{4.3}$$

$$SOC^{\mathrm{min},[\nu]} \leq SOC_{k+l}^{[\nu]} \leq SOC^{\mathrm{max},[\nu]} \tag{4.4}$$

with $\eta_{\mathrm{ch}}^{[\nu]}$ and $\eta_{\mathrm{d}}^{[\nu]}$ the charging and discharging efficiency of the battery. Moreover, T_{s} is the sampling time for the discretization. The physical boundaries of the battery storage are modeled by a maximum SOC^{max} and minimum SOC^{min} state of charge constraint.

Inverter capacity constraints

Every component in the nodes $\nu \in \mathcal{N}$ and $l \in \{0, \cdots, N_{\mathrm{p}} - 1\}$ that is interfaced to the grid with an inverter has a specific apparent power constraint which is given by the rated power of the inverter. The power output formulation can be written as a quadratic constraint in form of a squared two norm as

$$\left\| \begin{pmatrix} p_{\mathrm{ch},k+l}^{[\nu]} \\ q_{\mathrm{ch},k+l}^{[\nu]} \end{pmatrix} \right\|^2 \leq \left(S_{\mathrm{ch}}^{\max,[\nu]} \right)^2, \left\| \begin{pmatrix} p_{\mathrm{d},k+l}^{[\nu]} \\ q_{\mathrm{d},k+l}^{[\nu]} \end{pmatrix} \right\|^2 \leq \left(S_{\mathrm{d}}^{\max,[\nu]} \right)^2, \tag{4.5a}$$

$$\left\| \begin{pmatrix} p_{\mathrm{DG},k+l}^{[\nu]} \\ q_{\mathrm{DG},k+l}^{[\nu]} \end{pmatrix} \right\|^2 \leq \left(S_{\mathrm{DG}}^{\max,[\nu]} \right)^2, \left\| \begin{pmatrix} p_{\mathrm{EV},k+l}^{[\nu]} \\ q_{\mathrm{EV},k+l}^{[\nu]} \end{pmatrix} \right\|^2 \leq \left(S_{\mathrm{EV}}^{\max,[\nu]} \right)^2. \tag{4.5b}$$

Additionally, there are linear constraints on the reactive power $p_{\mathrm{ch},k+l}^{[\nu]}, p_{\mathrm{d},k+l}^{[\nu]}$, the battery storage inverters, the charging power of the electric vehicles $p_{\mathrm{EV},k+l}^{[\nu]}$ and the distributed generation $p_{\mathrm{DG},k+l}^{[\nu]}$.

$$p_{\mathrm{d},k+l}^{[\nu]} \leq 0,\ 0 \leq p_{\mathrm{ch},k+l}^{[\nu]} \tag{4.6a}$$

$$0 \leq p_{\mathrm{DG},k+l}^{[\nu]} \leq \hat{p}_{\mathrm{DG},k+l}^{[\nu]} \tag{4.6b}$$

$$0 \leq p_{\mathrm{EV},k+l}^{[\nu]} \leq \hat{p}_{\mathrm{EV},k+l}^{[\nu]} \tag{4.6c}$$

The inverter constraint (4.5a)-(4.5b) are summarized in one set as $\mathcal{S}_{\mathrm{loc},k+l}^{[\nu]}$

4.2.2 Distributed four step negotiation algorithm

Step 1: Performance chart construction:

Mathematical Minkowski operations on elliptical and polytopic sets are well established [FSR12]. The Minkowski sum of two convex sets centered at the origin, n_1 and n_2 in the planar space, is denoted by $\mathcal{F}_1 \oplus \mathcal{F}_2$ and defined as

$$\mathcal{F}_1 \oplus \mathcal{F}_2 = \{ f_1 + f_2 | f_1 \in \mathcal{F}_1, f_2 \in \mathcal{F}_2 \}. \tag{4.7}$$

Even though the definition looks simple, in general the operation tends to be very computational demanding, and not useful for on-line evaluations. However, for planar ellipsoids in $\mathbb{R}^2$ the parametric equation

$$\begin{pmatrix} x_2 \\ x_2 \end{pmatrix}^{\mathrm{T}} \mathcal{W}^{\mathrm{T}} \Lambda(\boldsymbol{a}) \mathcal{W} \begin{pmatrix} x_1 \\ x_2 \end{pmatrix} \leq 1 \tag{4.8}$$

can be used. With the matrix $\mathcal{W}$ describing the rotation and the diagonal matrix $\Lambda(\boldsymbol{a}) = \mathrm{diag}\,[a_1, a_2]$ defining the semi-axis length. It is assumed that operation (4.7) is only calculated to pairs of ellipsoids defined by (4.8) for which the pseudo-disk property [Fog11] applies. Furthermore, the investigated elliptical sets have rotation matrices equal to the unity matrix $\mathcal{W} = \boldsymbol{I}$ and thus the same fixed orientation with respect to their semi-axis length. The Minkowski sum of two ellipsoids with these properties can then be found by just adding their respective extreme points of each of their semi-axis in $\mathbb{R}^2$ as depicted in Figure 4.3.

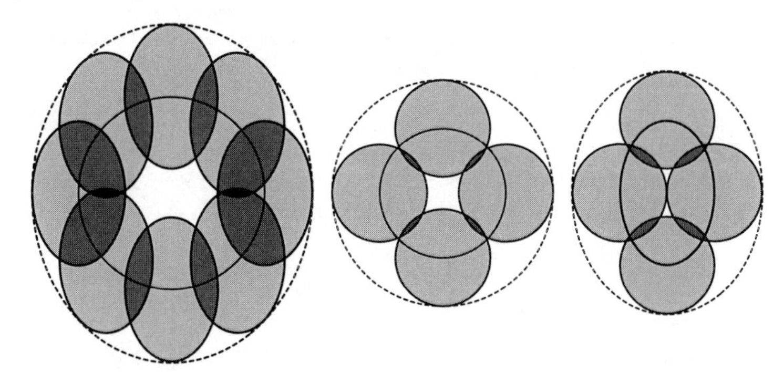

Figure 4.3: Visualization of the Minkowski sum for specific elliptical sets with fixed orientation and pseudo-disk property

The dotted outer set in Figure 4.3 depicts the sum of inner white and the grey colored ellipsoid. If the before mentioned conditions are met by the initial ellipsoids this sum breaks down to a simple addition of the respective extreme points of the semi-axis lengths. In the next sub-step, the constraint set construction based on these properties is described.

First each controller $\nu \in \mathcal{N}$ computes its node capability for the local exchange of apparent power $\boldsymbol{s}_{k+l}^{[\nu]} = [p_{k+l}^{[\nu]}, q_{k+l}^{[\nu]}]^{\mathrm{T}}$ with the grid for each step l in the prediction horizon N_{P} and broadcasts them to all others. The central idea behind this, is that the controllers only exchange the bounds on their flexibility and reactive power support. Thus, giving some degrees of freedom to other controllers for the negotiation.

The power limits for the chart construction are calculated locally based on (4.9)

$$\begin{aligned} P_{k+l}^{\mathrm{up},[\nu]} &= \hat{p}_{\mathrm{L},k+l}^{[\nu]} + \hat{p}_{\mathrm{DG},k+l}^{[\nu]} + P_{\mathrm{d},k+l}^{\mathrm{up},[\nu]} \\ P_{k+l}^{\mathrm{lo},[\nu]} &= \hat{p}_{\mathrm{L},k+l}^{[\nu]} - \hat{p}_{\mathrm{DG},k+l}^{[\nu]} + P_{\mathrm{ch},k+l}^{\mathrm{lo},[\nu]} \\ Q_{k+l}^{\mathrm{up},[\nu]} &= \hat{q}_{\mathrm{L},k+l}^{[\nu]} + Q_{\mathrm{DG}}^{\mathrm{up},[\nu]} + Q_{\mathrm{ch}}^{\mathrm{lo},[\nu]} \\ Q_{k+l}^{\mathrm{lo},[\nu]} &= \hat{q}_{\mathrm{L},k+l}^{[\nu]} - Q_{\mathrm{DG}}^{\mathrm{lo},[\nu]} - Q_{\mathrm{d}}^{\mathrm{up},[\nu]}. \end{aligned} \tag{4.9}$$

Note that the bounds follow from the power split (4.1) and (4.2). The load prediction $\hat{p}_{\mathrm{L},k+l}^{[\nu]}, \hat{q}_{\mathrm{L},k+l}^{[\nu]}$ is directly included as the load cannot be controlled. On the other hand, the bound on the photo-voltaic generation unit is based on the predicted value $\hat{p}^{[\nu]}_{\mathrm{DG},k+l}$, as the inverter output $p^{[\nu]}_{\mathrm{DG},k}$ can be adjusted up to this boundary. The reactive power components $Q^{\mathrm{lo},[\nu]}_{\mathrm{ch}}, Q^{\mathrm{up},[\nu]}_{\mathrm{d}}, Q^{\mathrm{up},[\nu]}_{\mathrm{DG}}, Q^{\mathrm{lo},[\nu]}_{\mathrm{DG}}$ are constructional features depending on the apparent power rating of the inverters. Note that it assumed here that inverters can provide reactive power even outside the in-feed hours [AYB15].

The bounds on maximum charge and discharge powers $P^{\mathrm{lo},[\nu]}_{\mathrm{ch},k}$ and $P^{\mathrm{up},[\nu]}_{\mathrm{d},k}$ are more difficult to choose. They should be as large as possible but must not exceed the maximum charge and discharge power $P^{\mathrm{min},[\nu]}_{\mathrm{ch}}$ and $P^{\mathrm{max},[\nu]}_{\mathrm{d}}$ of the battery inverter interface. Moreover, (4.3) and (4.4) have to be satisfied over the prediction horizon. This leads to

$$P^{\mathrm{lo},[\nu]}_{\mathrm{ch},k+l} = \max\left(P^{\mathrm{min},[\nu]}_{\mathrm{ch}}, \frac{1}{\eta^{[\nu]}_{\mathrm{ch}} T_{\mathrm{s}}}\left(SOC^{\mathrm{max},[\nu]} - SOC^{[\nu]}_k + \sum_{i=k}^{k+l-1} \eta^{[\nu]}_{\mathrm{ch}} T_{\mathrm{s}} P^{\mathrm{lo},[\nu]}_{\mathrm{ch},i}\right)\right)$$

$$P^{\mathrm{up},[\nu]}_{\mathrm{d},k+l} = \min\left(P^{\mathrm{max},[\nu]}_{\mathrm{d}}, \frac{\eta^{[\nu]}_{\mathrm{d}}}{T_{\mathrm{s}}}\left(SOC^{\mathrm{min},[\nu]} - SOC^{[\nu]}_k - \sum_{i=k}^{k+l-1} \frac{T_{\mathrm{s}}}{\eta^{[\nu]}_{\mathrm{d}}} P^{\mathrm{up},[\nu]}_{\mathrm{d},i}\right)\right)$$

for all $l \in \{0, \ldots, N_{\mathrm{p}}\}$. Noteworthy, $P^{\mathrm{lo},[\nu]}_{\mathrm{c},k+l}$ and $P^{\mathrm{up},[\nu]}_{\mathrm{d},k+l}$ are calculated recursively and can be interpreted as the bounds on the charging power such that in the worst case, i.e. when other local controller use power that has to be supplied by the storage of node ν, the energy constraints (4.4) are not violated.

Now the constraints on the apparent powers which are exchanged between the controllers can be formulated. They are defined by an ellipsoidal set of the following form

$$\mathcal{S}^{[\nu]}_{k+l} = \{\boldsymbol{s} \in \mathbb{R}^2 | (\boldsymbol{s} - \boldsymbol{s}_{0,k+l})^{\mathrm{T}} \boldsymbol{D}^{[\nu]}_{k+l} (\boldsymbol{s} - \boldsymbol{s}_{0,k+l}) \leq 1\}. \tag{4.10}$$

The matrix $\boldsymbol{D}^{[\nu]}_{k+l}$ together with the center $\boldsymbol{s}^{[\nu]}_{0,k+l}$ of the performance chart are given by

$$\boldsymbol{s}^{[\nu]}_{0,k+l} = \begin{pmatrix} \frac{P^{\mathrm{up},[\nu]}_{k+l} + P^{\mathrm{lo},[\nu]}_{k+l}}{2} \\ \frac{Q^{\mathrm{up},[\nu]}_{k+l} + Q^{\mathrm{lo},[\nu]}_{k+l}}{2} \end{pmatrix}, \boldsymbol{D}^{[\nu]}_{k+l} = \begin{pmatrix} \frac{1}{\left(r^{[\nu]}_{p,k+l}\right)^2} & 0 \\ 0 & \frac{1}{\left(r^{[\nu]}_{q,k+l}\right)^2} \end{pmatrix} \tag{4.11}$$

$$\text{with } r^{[\nu]}_{p,k+l} = \frac{P^{\mathrm{up},[\nu]}_{k+l} - P^{\mathrm{lo},[\nu]}_{k+l}}{2} \quad r^{[\nu]}_{q,k+l} = \frac{Q^{\mathrm{up},[\nu]}_{k+l} - Q^{\mathrm{lo},[\nu]}_{k+l}}{2}.$$

Based on these formulation a graphical representation for a simplified chart to visualize the result is shown in Figure 4.4. For this purpose the power chart from a PV plant (1) is added to the capability of a battery storage (2) for a single time instant. In a third step the local load is added to the resulting power chart from (2) to receive the final result in (3). The center of (3) is equal to the local load prediction. The performance charts as depicted in Figure 4.4, are generated for every time instance $k+l$ for each l

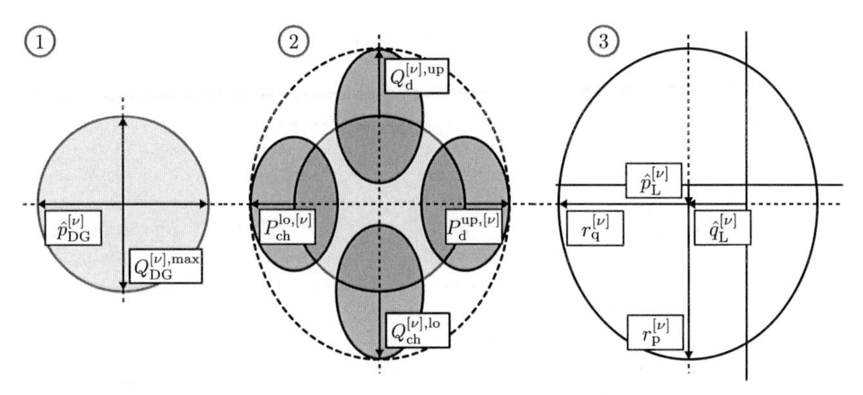

Figure 4.4: Visualization of the local performance chart construction based on (4.9) for one time instance k

in the prediction horizon N_{P}. After that they are broad-casted to every local controller $\nu \in \mathcal{N}$. The controllers then combine the individual node capabilities into a central one with $\mathcal{S}_{\mathrm{glob},k+l} = \mathcal{S}_{k+l}^{[1]} \times \mathcal{S}_{k+l}^{[2]} \times \cdots \times \mathcal{S}_{k+l}^{[N]}$ to formulate a global apparent power dispatch constraints.

Step 2: global optimal dispatch proposal

After each controller $[\nu]$ in the grid has received the information to generate the constraint $\mathcal{S}_{\mathrm{glob},k+l}$, it calculates a proposal for the global dispatch. By formulating a separable cost function in the form of $J_{\mathrm{glo},k}^{[\nu]} + J_{\mathrm{loc},k}^{[\nu]}$, global as well as local interest of each controller are considered. The global part of the cost function $J_{\mathrm{glo},k}^{[\nu]}$ can be written as following

$$\begin{aligned} J_{\mathrm{glo},k}^{[\nu]} =& \sum_{l=0}^{N_{\mathrm{p}}-1} C_{\mathrm{Loss}} \left(\boldsymbol{H}_{\mathrm{PF}} \boldsymbol{s}_{\mathrm{glo},k+l}^{[\nu]} \right)^{\mathrm{T}} \boldsymbol{R} \otimes \boldsymbol{I}^{2\times 2} \boldsymbol{H}_{\mathrm{PF}} \boldsymbol{s}_{\mathrm{glo},k+l}^{[\nu]} + \\ & \sum_{l=0}^{N_{\mathrm{p}}-1} C_{\mathrm{kWh}} \left[\left(\mathbf{1}^{\mathrm{T}} - \mathbf{1}_{\nu}^{\mathrm{T}} \right) \mathbf{0}^{\mathrm{T}} \right] \boldsymbol{s}_{\mathrm{glo},k+l}^{[\nu]} \end{aligned} \tag{4.12}$$

with

$$\boldsymbol{s}_{\mathrm{glo},k+l}^{[\nu]} = \left[p_{k+l}^{[1]}, p_{k+l}^{[2]}, \cdots, p_{k+l}^{[N]}, q_{k+l}^{[1]}, q_{k+l}^{[2]}, \cdots, q_{k+l}^{[N]} \right]^{\mathrm{T}}, \forall l \in \{0, ..., N_{\mathrm{p}} - 1\}.$$

In most loss formulations for an OPF either the branch currents or the nodal voltages are used to express losses [CGW05, BCCZ15]. With the use of the developed LIPF the

nodal currents can be expressed as a linear function of the apparent power injections of each node as $\boldsymbol{H}_{\mathrm{PF}} = -\boldsymbol{A}_{\mathrm{PF}}^{-1}\boldsymbol{B}_{\mathrm{PF}}$, with $\boldsymbol{A}_{\mathrm{PF}}$ and $\boldsymbol{B}_{\mathrm{PF}}$ from (2.30).
The resistance matrix $\boldsymbol{R}$ inherits the topology information from the incidence matrix, so only the Kronecker product $\boldsymbol{R}\otimes\boldsymbol{I}^{2\times2}$ is needed to expanded the matrix to an appropriate size, where $\boldsymbol{I}^{2\times2}$ is the identity matrix in with size 2×2. If the branch currents would be used instead to calculate losses, the resistance matrix would have the branch resistances on the diagonal.
Consequently the first summand represents the costs for grid losses. The second summand of the objective represents the approximated benefit of active power dispatch from other controllers. As has been noted in the introduction, it is based on the assumption, that domestic owners want to in-feed power from PV into the grid and at the same time minimize their power consumption. However, it is important that the controllers exclude themselves from the approximated part in their own proposal, as this would distort the result. By setting the appropriate local cost approximation to zero with $\mathbf{1}^{\mathrm{T}} - \mathbf{1}_{\nu}^{\mathrm{T}}$ this can be achieved. Moreover, the global apparent power vector $\boldsymbol{s}_{\mathrm{glo},l}^{[\nu]}$ represents the grid interaction variables on the left hand side of the equality sign called local power split (4.1) + (4.2) of every node ν as depicted in Figure 4.2. The terms C_{Loss}, C_{kWh} are assumed to be known to every controller and represent together with the sample time of the discrete-time system, a fixed energy price.

Next the controller has to recognize the local interest of the area or node it operates. Therefore, the optimization problem has to consider a local part of the objective represented as a linear function, that reflects the interests of the domestic owner. It is formulated as

$$J_{\mathrm{loc},k}^{[\nu]} = \sum_{l=0}^{N_{\mathrm{p}}-1} \begin{pmatrix} C_{\mathrm{DG}}^{[\nu]} & C_{\mathrm{EV}}^{[\nu]} & C_{\mathrm{ch}}^{[\nu]} & C_{\mathrm{d}}^{[\nu]} & 0 & 0 & 0 & 0 \end{pmatrix} \boldsymbol{s}_{\mathrm{loc},k+l}^{[\nu]} \tag{4.13}$$

with

$$\boldsymbol{s}_{\mathrm{loc},k+l}^{[\nu]} = \left[p_{\mathrm{DG},k+l}^{[\nu]}, p_{\mathrm{EV},k+l}^{[\nu]}, p_{\mathrm{ch},k+l}^{[\nu]}, p_{\mathrm{d},k+l}^{[\nu]}, q_{\mathrm{DG},k+l}^{[\nu]}, q_{\mathrm{EV},k+l}^{[\nu]}, q_{\mathrm{ch},k+l}^{[\nu]}, q_{\mathrm{d},k+l}^{[\nu]}\right]^{\mathrm{T}}, \forall l \in \{0, ..., N_{\mathrm{p}}-1\}$$

where $C_{\mathrm{DG}}^{[\nu]}$ is the benefit from providing active power from the DG in form of an in-feed tariff. The terms $C_{\mathrm{ch}}^{[\nu]}$ and $C_{\mathrm{d}}^{[\nu]}$ representing the wear costs when using the local battery storage. $C_{\mathrm{EV}}^{[\nu]}$ is a costumer benefit for charging its electric vehicle with as much power as possible. The optimization variable comprises of all variables on the right hand side of (4.1) + (4.2). For the local part it only considers the variables of node ν and not of the other nodes, as in the global part.
From the form of the objectives it is clear that this is a multi-objective optimization problem which is the usual case for reactive power support or voltage control applications [MDS+17].

To ensure a secure operation, respecting the boundaries of the grid assets and components in the power system, constraints have to be introduced. The grid wide constraints for $\nu, \mu \in \mathcal{N}$and $l \in \mathcal{P} = \{0, \dots, N_\mathrm{P} - 1\}$ are derived first.

$$\boldsymbol{s}_{\mathrm{glo},k+l}^{[\nu]} \in \boldsymbol{S}_{\mathrm{glo},k+l} \tag{4.14a}$$

$$\boldsymbol{s}_{\mathrm{glo},k}^{[\nu],\mathrm{T}} \boldsymbol{H}_{\mathrm{PF}}^{\mathrm{T}} \begin{pmatrix} \tilde{\boldsymbol{W}}^{-1} & \boldsymbol{0} \\ \boldsymbol{0} & \tilde{\boldsymbol{W}}^{-1} \end{pmatrix}^{\mathrm{T}} \boldsymbol{2}_{\iota} \boldsymbol{I}^{2\times 2} \boldsymbol{2}_{\iota}^{\mathrm{T}} \begin{pmatrix} \tilde{\boldsymbol{W}}^{-1} & \boldsymbol{0} \\ \boldsymbol{0} & \tilde{\boldsymbol{W}}^{-1} \end{pmatrix} \boldsymbol{H}_{\mathrm{PF}} \boldsymbol{s}_{\mathrm{glo},k+l}^{[\nu]} \leq I_{\mathrm{Branch}}^{\mathrm{up},[\iota]} \tag{4.14b}$$

with $\boldsymbol{2}_{\iota}^{\mathrm{T}} = \begin{pmatrix} \boldsymbol{1}_{\iota}^{\mathrm{T}} & \boldsymbol{0} \\ \boldsymbol{0} & \boldsymbol{1}_{\iota}^{\mathrm{T}} \end{pmatrix}$ a matrix that extracts the real and imaginary part of the branch current ι and $\tilde{\boldsymbol{W}}^{-1}$ a sub-inverse of the incidence matrix $\boldsymbol{W}$. This matrix is calculated by eliminating the first column of $\boldsymbol{W}$ and then inverting the resulting sub-matrix. $I_{\mathrm{Branch}}^{\mathrm{up},[\iota]}$ represents the current-carrying capacity of the low voltage underground cables for each branch $\iota \in \mathcal{N}_{-0}$, where $\mathcal{N}_{-0}$ means that in a radial grid there are N nodes but $N-1$ branches.

The global apparent power dispatch constraint $\boldsymbol{S}_{\mathrm{glo},k+l}$ directly follows from the power chart construction. Usually the voltage magnitude constraint would be formulated as

$$U_{\mathrm{Node}}^{\mathrm{lo},[\nu]} \leq \left\| \boldsymbol{2}_{\nu}^{\mathrm{T}} \boldsymbol{C}_{\mathrm{PF}} \boldsymbol{H}_{\mathrm{PF}} \boldsymbol{s}_{\mathrm{glo},k+l}^{[\nu]} + \begin{pmatrix} u_0 \\ 0 \end{pmatrix} \right\| \leq U_{\mathrm{Node}}^{\mathrm{up},[\nu]}. \tag{4.15}$$

With $U_{\mathrm{Node}}^{\mathrm{up},[\nu]}$ the upper and $U_{\mathrm{Node}}^{\mathrm{lo},[\nu]}$ the lower bound on the nodal voltage magnitude according to [VDE18]. Especially the nodal voltage constraints have to be discussed further, because the lower bound is non-convex. In the following part a relaxation is introduced to handle both bounds with a single constraint.

Relaxation of the non-convex voltage constraints

The lower limit of the voltage constraint (4.15) is non-convex. In literature the proposed solution to this problem is usually an inner approximation of the elliptic voltage constraint by means of a polytopic approximation [For17]. But this approach has the drawback that it results in a reduced representation of the inverter grid interaction for voltage control. This is mainly related to the fact that consumed reactive power q_k, to mitigate the voltage magnitude rise, changes the phase angle. Especially in LV grids with the low X/R ratio, this effect is strong. When using polytopes to formulate the voltage constraint, it is clearly visible, that the voltage $\underline{u}_{\mathrm{poly}}$ needs to be reduced to consume reactive power. In case of an elliptic approximation, the voltage magnitude of $\underline{u}_{\mathrm{ellip}}$ stays constant and only the angle changes. The difference between both formulations is $\Delta\underline{u}$. Both situations are depicted in Figure 4.5 on the right hand side. The behavior with polytopic constraints does not reflect the reality. It would mean that the inverters have to consume additional reactive power, which increases grid losses. If this cannot

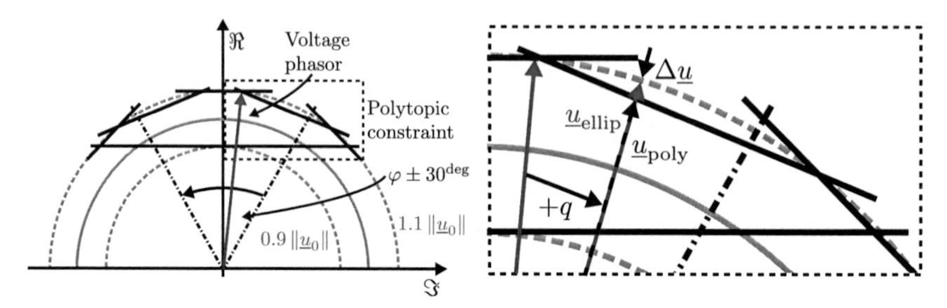

Figure 4.5: Inner approximation of the elliptic voltage constraints by linear polytopes, with a maximum phase angle φ of 30°

be achieved, they would need to reduce active power in-feed. As a result the inverters will not react the way they suppose to when using polytopic constraints. Because the PV units in the developed approach are not modeled as negative loads but instead have a curtailment functionality related to a cost in the objective function, this effect is not tolerable. Instead, a combination of box constraints and an ellipsoidal constraint is used to approximate the real voltage constraints. One major advantage of this approach is that the upper constraint of the voltage is approximated with high precision and thus the behavior of the local controllers are much closer to reality.

Another factor is the lower voltage constraint, which would be non-convex for the case in (4.15). An elliptic formulation in combination with a linear constraint can cover this point as well, as depicted in Figure 4.6. The Ellipsoidal constraint can be formulated as

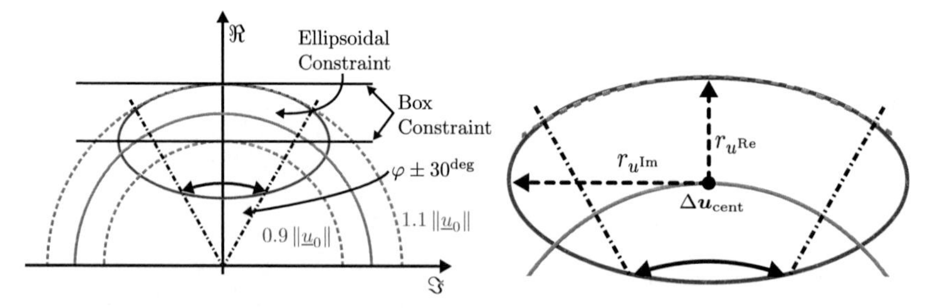

Figure 4.6: Approximation of the voltage limitation by the combination of an elliptic and polytopic constraint

$$\underbrace{\boldsymbol{s}_{\mathrm{glo},k}^{[\nu],\mathrm{T}} \boldsymbol{H}_{\mathrm{PF}}^{\mathrm{T}} \boldsymbol{C}_{\mathrm{PF}}^{\mathrm{T}} \boldsymbol{2}_{\nu} \begin{pmatrix} r_{u\mathrm{Re}} & 0 \\ 0 & r_{u\mathrm{Im}} \end{pmatrix} \boldsymbol{2}_{\nu}^{\mathrm{T}} \boldsymbol{C}_{\mathrm{PF}} \boldsymbol{H}_{\mathrm{PF}} \boldsymbol{s}_{\mathrm{glo},k}^{[\nu]}}_{\boldsymbol{O}_k^{[\nu]}} \\ \underbrace{-2\Delta \boldsymbol{u}_{\mathrm{cent}}^{\mathrm{T}} \boldsymbol{C}_{\mathrm{PF}} \boldsymbol{H}_{\mathrm{PF}}}_{\boldsymbol{l}_k^{[\nu]}} \boldsymbol{s}_{\mathrm{glo},k}^{[\nu]} \leq U_{\mathrm{Node}}^{\mathrm{up},[\nu]} - \|\underline{u}_0\| \,. \tag{4.16}$$

The determination of the radii $r_{u\mathrm{Re}}$, $r_{u\mathrm{Im}}$ and the center $\Delta \boldsymbol{u}_{\mathrm{cent}}$ of the ellipsoid is straight forward. By choosing the center to $[0.9, 0]^{\mathrm{T}}$ and the radius $r_{u\mathrm{Re}} = 0.2$, $r_{u\mathrm{Im}}$ can be determined by maximizing the semi-axis until there is collision between the outer voltage ellipse with $1.1\,\|\underline{u}_0\|$. This procedure has been described in [Var06, Nie16]. Furthermore, the polytopic constraints follow as

$$\begin{pmatrix} \boldsymbol{C}_{\mathrm{PF}} \boldsymbol{H}_{\mathrm{PF}} \\ -\boldsymbol{C}_{\mathrm{PF}} \boldsymbol{H}_{\mathrm{PF}} \end{pmatrix} \boldsymbol{s}_{\mathrm{glo},k}^{[\nu]} \leq \begin{pmatrix} U_{\mathrm{Node}}^{\mathrm{up},[\nu]} - \|\underline{u}_0\| \\ U_{\mathrm{Node}}^{\mathrm{lo},[\nu]} - \|\underline{u}_0\| \end{pmatrix} . \tag{4.17}$$

The lost area with this approach, is one of high phase angle φ_ν at the lower voltage bound which is very uncommon. It would correspond to a high inductive power in-feed at a high loading. This can happen, but is not the standard situation for reactive power support in the grid. After setting up the system-wide constraints, every local controller can calculate a dispatch solution that satisfies the limitations of the grid equipment and the flexibility of other sub-systems. In the second step, the constraints that complement the local objective are introduced.

$$\boldsymbol{s}_{\mathrm{loc},k+l}^{[\nu]} \in \mathcal{S}_{\mathrm{loc},k+l}^{[\nu]} \tag{4.18a}$$

$$\begin{pmatrix} p_{k+l}^{[\nu]} \\ q_{k+l}^{[\nu]} \end{pmatrix} - \begin{pmatrix} \begin{pmatrix} 1 & 1 & 1 & 1 \end{pmatrix} & \boldsymbol{0} \\ \boldsymbol{0} & \begin{pmatrix} 1 & 1 & 1 & 1 \end{pmatrix} \end{pmatrix} \boldsymbol{s}_{\mathrm{loc},k+l}^{[\nu]} = \begin{pmatrix} \hat{p}_{\mathrm{L},k+l}^{[\nu]} \\ \hat{q}_{\mathrm{L},k+l}^{[\nu]} \end{pmatrix} \tag{4.18b}$$

$$\begin{pmatrix} \hat{p}_{\mathrm{L},k}^{[\nu]} \\ \hat{q}_{\mathrm{L},k}^{[\nu]} \end{pmatrix} = \begin{pmatrix} \overline{p}_{\mathrm{L},k}^{[\nu]} \\ \overline{q}_{\mathrm{L},k}^{[\nu]} \end{pmatrix} \tag{4.18c}$$

$$SOC_{k+l+1}^{[\nu]} = SOC_{k+l}^{[\nu]} + \underbrace{\begin{pmatrix} \boldsymbol{0} & T_{\mathrm{s}}\eta_{\mathrm{ch}} & T_{\mathrm{s}}\frac{1}{\eta_{\mathrm{d}}} & \boldsymbol{0} \end{pmatrix}}_{\boldsymbol{G}} \boldsymbol{s}_{\mathrm{loc},k+l}^{[\nu]} \tag{4.18d}$$

$$SOC_k^{[\nu]} = \overline{SOC}_k^{[\nu]} \tag{4.18e}$$

The first constraint (4.18a) is given by the inverter ratings introduced in (4.6c)-(4.5a). Constraint (4.18b) is responsible to keep the power balance (4.1) and (4.2) and at the same time force the load prediction $\hat{p}_{\mathrm{L},k+l}^{[\nu]}, \hat{q}_{\mathrm{L},k+l}^{[\nu]}$ to be held. Furthermore, the load prediction is never exact, but to limit the error for the calculation, the first predicted value is forced onto the last measured load power values $\overline{p}_{\mathrm{L},k}^{[\nu]}, \overline{q}_{\mathrm{L},k}^{[\nu]}$. The battery storage is considered with dynamics (4.18d), battery state of charge (4.4) as well as persistence constraints (4.18e) with regard to the measured state of charge.

The inequality constraints are integrated in the optimization problem by combining the prediction model in (1.4) with the inequality formulations. This way the state of charge is not visible as an optimization variable. The inequality constraints for one time instance k are formulated as

$$\underbrace{\begin{bmatrix} -1 \\ 1 \\ \mathbf{0} \\ \mathbf{0} \\ \mathbf{0} \\ \mathbf{0} \\ \mathbf{0} \\ \mathbf{0} \end{bmatrix}}_{\boldsymbol{M}_{\mathrm{glo},k}} SOC_k^{[\nu]} + \underbrace{\begin{bmatrix} \mathbf{0} & -\boldsymbol{G} \\ \mathbf{0} & \boldsymbol{G} \\ -\boldsymbol{C}_{\mathrm{PF}}\boldsymbol{H}_{\mathrm{PF}} & \mathbf{0} \\ \boldsymbol{C}_{\mathrm{PF}}\boldsymbol{H}_{\mathrm{PF}} & \mathbf{0} \\ \mathbf{0} & \boldsymbol{A}_{\mathrm{loc}}^{\mathrm{min},p} \\ \mathbf{0} & \boldsymbol{A}_{\mathrm{loc}}^{\mathrm{max},p} \\ \mathbf{0} & \boldsymbol{A}_{\mathrm{loc}}^{\mathrm{min},q} \\ \mathbf{0} & \boldsymbol{A}_{\mathrm{loc}}^{\mathrm{max},q} \end{bmatrix}}_{\boldsymbol{H}_{\mathrm{glo},k}} \begin{pmatrix} \boldsymbol{s}_{\mathrm{glob},k}^{[\nu]} \\ \boldsymbol{s}_{\mathrm{loc},k}^{[\nu]} \end{pmatrix} + \underbrace{\begin{bmatrix} 0 \\ 0 \\ -\boldsymbol{D}_{\mathrm{PF}} \\ \boldsymbol{D}_{\mathrm{PF}} \\ \mathbf{0} \\ \mathbf{0} \\ \mathbf{0} \\ \mathbf{0} \end{bmatrix}}_{\boldsymbol{V}_{\mathrm{glo},k}} \boldsymbol{u}_0 \leq \underbrace{\begin{bmatrix} -SOC^{\mathrm{min},[\nu]} \\ SOC^{\mathrm{max},[\nu]} \\ -\Delta\boldsymbol{u}_{\mathrm{Node}} \\ \Delta\boldsymbol{u}_{\mathrm{Node}} \\ -\boldsymbol{p}_k^{\mathrm{min},[\nu]} \\ \boldsymbol{p}_k^{\mathrm{max},[\nu]} \\ -\boldsymbol{q}_k^{\mathrm{min},[\nu]} \\ \boldsymbol{q}_k^{\mathrm{max},[\nu]} \end{bmatrix}}_{\boldsymbol{b}_{\mathrm{glo},k}} . \tag{4.19}$$

With $\boldsymbol{A}_{\mathrm{loc}}^{\mathrm{min},p}, \boldsymbol{A}_{\mathrm{loc}}^{\mathrm{max},p}, \boldsymbol{A}_{\mathrm{loc}}^{\mathrm{min},q}, \boldsymbol{A}_{\mathrm{loc}}^{\mathrm{max},q}$ being appropriate matrices which map the local inequalities onto its set of admissible variables $\boldsymbol{p}^{\mathrm{max},[\nu]}, \boldsymbol{p}^{\mathrm{min},[\nu]}, \boldsymbol{q}^{\mathrm{max},[\nu]}, \boldsymbol{q}^{\mathrm{min},[\nu]}$ which are defined by (4.6c)-(4.6a). $\Delta\boldsymbol{u}_{\mathrm{Node}}$ is a fixed vector of voltage magnitude difference between nominal $\|\underline{u}_0\|$ and the symmetric boundary $U_{\mathrm{Node}}^{\mathrm{lo(up)},[\nu]}$ introduced in (4.17).

Rewriting the inequality constraint for the full prediction horizon N_{p} and including the optimizations variables as well as the state of charge of the battery storage leads to

$$\underbrace{\begin{bmatrix} \boldsymbol{M}_{\mathrm{glo},k} \\ \mathbf{0} \\ \vdots \\ \mathbf{0} \end{bmatrix}}_{\boldsymbol{\mathcal{D}}_{\mathrm{glo}}} \overline{SOC}_k^{[\nu]} + \underbrace{\begin{bmatrix} \mathbf{0} & \cdots & \mathbf{0} \\ \boldsymbol{M}_{\mathrm{glo},k+1} & \cdots & \mathbf{0} \\ \vdots & \ddots & \vdots \\ \mathbf{0} & \cdots & \boldsymbol{M}_{\mathrm{glo},k+N_{\mathrm{p}}} \end{bmatrix}}_{\boldsymbol{\mathcal{M}}_{\mathrm{glo}}} \underbrace{\begin{bmatrix} SOC_{k+1}^{[\nu]} \\ SOC_{k+2}^{[\nu]} \\ \vdots \\ SOC_{k+N_{\mathrm{p}}}^{[\nu]} \end{bmatrix}}_{\boldsymbol{SOC}^{[\nu]}} +$$

$$\underbrace{\begin{bmatrix} \boldsymbol{V}_{\mathrm{glo},k} & \cdots & \mathbf{0} \\ \vdots & \ddots & \vdots \\ \mathbf{0} & \cdots & \boldsymbol{V}_{\mathrm{glo},k+N_{\mathrm{p}}-1} \\ \mathbf{0} & \cdots & \mathbf{0} \end{bmatrix}}_{\boldsymbol{\mathcal{E}}_{\mathrm{glo}}} \underbrace{\begin{bmatrix} \left[\boldsymbol{s}_{\mathrm{glo},k}^{[\nu],\mathrm{T}} \; \boldsymbol{s}_{\mathrm{loc},k}^{[\nu],\mathrm{T}}\right]^{\mathrm{T}} \\ \left[\boldsymbol{s}_{\mathrm{glo},k+1}^{[\nu],\mathrm{T}} \; \boldsymbol{s}_{\mathrm{loc},k+1}^{[\nu],\mathrm{T}}\right]^{\mathrm{T}} \\ \vdots \\ \left[\boldsymbol{s}_{\mathrm{glo},k+N_{\mathrm{p}}-1}^{[\nu],\mathrm{T}} \; \boldsymbol{s}_{\mathrm{loc},k+N_{\mathrm{p}}-1}^{[\nu],\mathrm{T}}\right]^{\mathrm{T}} \end{bmatrix}}_{\boldsymbol{S}_{\mathrm{glo}}^{[\nu]}} +$$

$$\underbrace{\begin{bmatrix} \boldsymbol{H}_{\mathrm{glo},k} & \cdots & 0 \\ \vdots & \ddots & \vdots \\ 0 & \cdots & \boldsymbol{H}_{\mathrm{glo},k+N_\mathrm{p}-1} \\ 0 & \cdots & 0 \end{bmatrix}}_{\boldsymbol{\mathcal{H}}_\mathrm{glo}} \underbrace{\begin{bmatrix} \boldsymbol{u}_0 \\ \boldsymbol{u}_0 \\ \vdots \\ \boldsymbol{u}_0 \end{bmatrix}}_{\boldsymbol{U}_0^{[\nu]}} \leq \underbrace{\begin{bmatrix} \boldsymbol{b}_{\mathrm{glo},k} \\ \boldsymbol{b}_{\mathrm{glo},k+1} \\ \vdots \\ \boldsymbol{b}_{\mathrm{glo},k+N_\mathrm{p}-1} \end{bmatrix}}_{\boldsymbol{B}_\mathrm{glo}}. \tag{4.20}$$

Which can be rewritten in compact form as

$$\boldsymbol{\mathcal{D}}_\mathrm{glo}\overline{SOC}_k^{[\nu]} + \boldsymbol{\mathcal{M}}_\mathrm{glo}\boldsymbol{SOC}^{[\nu]} + \boldsymbol{\mathcal{E}}_\mathrm{glo}\boldsymbol{S}_\mathrm{glo}^{[\nu]} + \boldsymbol{\mathcal{H}}_\mathrm{glo}\boldsymbol{U}_0^{[\nu]} \leq \boldsymbol{B}_\mathrm{glo}. \tag{4.21}$$

With the help of the prediction matrices for the battery storage $A_d = 1$, $\boldsymbol{B}_d = \boldsymbol{G}$, $\boldsymbol{C}_d = \boldsymbol{0}$ and $\boldsymbol{D}_d = \boldsymbol{0}$, equation (1.4) can be combined with (4.21) to

$$\boldsymbol{\mathcal{D}}_\mathrm{glo}\overline{SOC}_k^{[\nu]} + \boldsymbol{\mathcal{M}}_\mathrm{glo}\left(\boldsymbol{\Phi}\overline{SOC}_k^{[\nu]} + \boldsymbol{\Gamma}\boldsymbol{S}_\mathrm{glo}^{[\nu]}\right) + \boldsymbol{\mathcal{E}}_\mathrm{glo}\boldsymbol{S}^{[\nu]} + \boldsymbol{\mathcal{H}}_\mathrm{glo}\boldsymbol{U}_0^{[\nu]} \leq \boldsymbol{B}_\mathrm{glo} \tag{4.22a}$$

$$\left(\boldsymbol{\mathcal{D}}_\mathrm{glo} + \boldsymbol{\mathcal{M}}_\mathrm{glo}\boldsymbol{\Phi}\right)\overline{SOC}_k^{[\nu]} + \left(\boldsymbol{\mathcal{M}}_\mathrm{glo}\boldsymbol{\Gamma} + \boldsymbol{\mathcal{E}}_\mathrm{glo}\right)\boldsymbol{S}_\mathrm{glo}^{[\nu]} + \boldsymbol{\mathcal{H}}_\mathrm{glo}\boldsymbol{U}_0^{[\nu]} \leq \boldsymbol{B}_\mathrm{glo} \tag{4.22b}$$

$$\left(\boldsymbol{\mathcal{M}}_\mathrm{glo}\boldsymbol{\Gamma} + \boldsymbol{\mathcal{E}}_\mathrm{glo}\right)\boldsymbol{S}_\mathrm{glo}^{[\nu]} \leq \boldsymbol{B}_\mathrm{glo} - \left(\boldsymbol{\mathcal{D}}_\mathrm{glo} + \boldsymbol{\mathcal{M}}_\mathrm{glo}\boldsymbol{\Phi}\right)\overline{SOC}_k^{[\nu]} - \boldsymbol{\mathcal{H}}_\mathrm{glo}\boldsymbol{U}_0^{[\nu]}. \tag{4.22c}$$

After the introduction and generation of the constraints the optimization problem is solved with regard to a global $\boldsymbol{s}_{\mathrm{glo},k}^{[\nu]}$ and local $\boldsymbol{s}_{\mathrm{loc},k}^{[\nu]}$ optimization vector.

$$\underset{\boldsymbol{s}_{\mathrm{glo},k}^{[\nu]},\boldsymbol{s}_{\mathrm{loc},k}^{[\nu]}}{\text{minimize}} \quad J_{\mathrm{glo},k}^{[\nu]} + J_{\mathrm{loc},k}^{[\nu]}$$

subject to

$$\boldsymbol{s}_{\mathrm{loc},k+l}^{[\nu]} \in \mathcal{S}_{\mathrm{loc},k+l}^{[\nu]}$$

$$\boldsymbol{s}_{\mathrm{glo},k+l}^{[\nu]} \in \mathcal{S}_{\mathrm{glo},k+l}^{[\nu]}$$

$$\boldsymbol{s}_{\mathrm{glo},k}^{[\nu],\mathrm{T}}\boldsymbol{H}_\mathrm{PF}^\mathrm{T}\boldsymbol{C}_\mathrm{PF}^\mathrm{T}\boldsymbol{2}_\nu\begin{pmatrix} r_{u^\mathrm{Re}} & 0 \\ 0 & r_{u^\mathrm{Im}} \end{pmatrix}\boldsymbol{2}_\nu^\mathrm{T}\boldsymbol{C}_\mathrm{PF}\boldsymbol{H}_\mathrm{PF}\boldsymbol{s}_{\mathrm{glo},k}^{[\nu]} -$$

$$2\Delta\boldsymbol{u}_\mathrm{cent}^\mathrm{T}\boldsymbol{C}_\mathrm{PF}\boldsymbol{H}_\mathrm{PF}\boldsymbol{s}_{\mathrm{glo},k}^{[\nu]} \leq U_\mathrm{Node}^{\mathrm{up},[\nu]} - \|\underline{u}_0\|$$

$$\boldsymbol{s}_{\mathrm{glo},k}^{[\nu],\mathrm{T}}\boldsymbol{H}_\mathrm{PF}^\mathrm{T}\begin{pmatrix} \tilde{\boldsymbol{W}}^{-1} & \boldsymbol{0} \\ \boldsymbol{0} & \tilde{\boldsymbol{W}}^{-1} \end{pmatrix}^\mathrm{T}\boldsymbol{2}_\iota\boldsymbol{I}^{2\times 2}\boldsymbol{2}_\iota^\mathrm{T}\begin{pmatrix} \tilde{\boldsymbol{W}}^{-1} & \boldsymbol{0} \\ \boldsymbol{0} & \tilde{\boldsymbol{W}}^{-1} \end{pmatrix}\boldsymbol{H}_\mathrm{PF}\boldsymbol{s}_{\mathrm{glo},k+l}^{[\nu]} \leq I_\mathrm{Branch}^{\mathrm{up},[\iota]}$$

$$\begin{pmatrix} p_{k+l}^{[\nu]} \\ q_{k+l}^{[\nu]} \end{pmatrix} - \begin{pmatrix} \begin{pmatrix} 1 & 1 & 1 & 1 \end{pmatrix} & \boldsymbol{0} \\ \boldsymbol{0} & \begin{pmatrix} 1 & 1 & 1 & 1 \end{pmatrix} \end{pmatrix}\boldsymbol{s}_{\mathrm{loc},k+l}^{[\nu]} = \begin{pmatrix} \hat{p}_{\mathrm{L},k+l}^{[\nu]} \\ \hat{q}_{\mathrm{L},k+l}^{[\nu]} \end{pmatrix}$$

$$\begin{pmatrix} \hat{p}_{\mathrm{L},k}^{[\nu]} \\ \hat{q}_{\mathrm{L},k}^{[\nu]} \end{pmatrix} = \begin{pmatrix} \overline{p}_{\mathrm{L},k}^{[\nu]} \\ \overline{q}_{\mathrm{L},k}^{[\nu]} \end{pmatrix}$$

$$\left(\boldsymbol{\mathcal{M}}_\mathrm{glo}\boldsymbol{\Gamma} + \boldsymbol{\mathcal{E}}_\mathrm{glo}\right)\boldsymbol{S}_\mathrm{glo}^{[\nu]} \leq \boldsymbol{B}_\mathrm{glo} - \left(\boldsymbol{\mathcal{D}}_\mathrm{glo} + \boldsymbol{\mathcal{M}}_\mathrm{glo}\boldsymbol{\Phi}\right)\overline{SOC}_k^{[\nu]} - \boldsymbol{\mathcal{H}}_\mathrm{glo}\boldsymbol{U}^{[\nu]}$$

The resulting global part $\boldsymbol{s}_{\mathrm{glo},k+l}^{[\nu]}$ is then broadcast to every controller for $\kappa \in \mathcal{N}$ as a proposal for the global power dispatch. Keep in mind that this is done for every time step $l \in \mathcal{P} = \{0, \dots, N_\mathrm{P} - 1\}$, so that there is a planned dispatch for several steps into the future.

Step 3: Evaluation of other global dispatch proposals

In the last step every local controller has determined a solution for a global dispatch problem. For each proposal of active and reactive node power $p_k^{[\nu](\kappa)}$ and $q_k^{[\nu](\kappa)}$ received from controller (κ) (i.e. extracted from the broadcast variable $\boldsymbol{s}_{\mathrm{glo},k}^{(\kappa)}$), each local controller $[\nu]$ now computes its own optimal local power distribution such that (4.1) and (4.2) are satisfied. This is achieved by introducing an additional equality constraint that forces the local constraints on the dispatch solution of the other controllers as

$$\boldsymbol{2}_\nu^\mathrm{T} \boldsymbol{s}_{\mathrm{glo},k+l}^{(\kappa)} = \begin{pmatrix} (1\ \ 1\ \ 1\ \ 1) & \boldsymbol{0} \\ \boldsymbol{0} & (1\ \ 1\ \ 1\ \ 1) \end{pmatrix} \boldsymbol{s}_{\mathrm{loc},k+l}^{[\nu](\kappa)} + \begin{pmatrix} \hat{p}_{\mathrm{L},k+l}^{[\nu]} \\ \hat{q}_{\mathrm{L},k+l}^{[\nu]} \end{pmatrix}. \tag{4.23}$$

Recall that the superscript (κ) indicates that the solution is based on the proposed power dispatch of another controller (κ).

For this purpose first the local inequality constraints for the reevaluation are formulated as

$$\underbrace{\begin{bmatrix} -1 \\ 1 \\ 0 \\ 0 \\ 0 \\ 0 \end{bmatrix}}_{\boldsymbol{M}_{\mathrm{loc},k}} SOC_k^{[\nu]} + \underbrace{\begin{bmatrix} -\boldsymbol{G} \\ \boldsymbol{G} \\ \boldsymbol{A}_\mathrm{loc}^{\min,p} \\ \boldsymbol{A}_\mathrm{loc}^{\max,p} \\ \boldsymbol{A}_\mathrm{loc}^{\min,q} \\ \boldsymbol{A}_\mathrm{loc}^{\max,q} \end{bmatrix}}_{\boldsymbol{H}_{\mathrm{loc},k}} \boldsymbol{s}_{\mathrm{loc},k}^{[\nu](\kappa)} + \underbrace{\begin{bmatrix} -SOC^{\min,[\nu]} \\ SOC^{\max,[\nu]} \\ -\boldsymbol{p}_k^{\min,[\nu]} \\ \boldsymbol{p}_k^{\max,[\nu]} \\ -\boldsymbol{q}_k^{\min,[\nu]} \\ \boldsymbol{q}_k^{\max,[\nu]} \end{bmatrix}}_{\boldsymbol{b}_{\mathrm{loc},k}} . \tag{4.24}$$

Now again they are expanded for the prediction horizon

$$\underbrace{\begin{bmatrix} \boldsymbol{M}_{\mathrm{loc},k} \\ \boldsymbol{0} \\ \vdots \\ \boldsymbol{0} \end{bmatrix}}_{\boldsymbol{\mathcal{D}}_\mathrm{loc}} \overline{SOC}_k^{[\kappa]} + \underbrace{\begin{bmatrix} \boldsymbol{0} & \cdots & \boldsymbol{0} \\ \boldsymbol{M}_{\mathrm{loc},k+1} & \cdots & \boldsymbol{0} \\ \vdots & \ddots & \vdots \\ \boldsymbol{0} & \cdots & \boldsymbol{M}_{\mathrm{loc},k+N_\mathrm{p}-1} \end{bmatrix}}_{\boldsymbol{\mathcal{M}}_\mathrm{loc}} \underbrace{\begin{bmatrix} SOC_{k+1}^{[\nu]} \\ SOC_{k+2}^{[\nu]} \\ \vdots \\ SOC_{k+N_\mathrm{p}-1}^{[\nu]} \end{bmatrix}}_{\boldsymbol{SOC}^{[\kappa]}} +$$

$$\underbrace{\begin{bmatrix} \boldsymbol{V}_{\text{loc},k} & \cdots & \mathbf{0} \\ \vdots & \ddots & \vdots \\ \mathbf{0} & \cdots & \boldsymbol{V}_{\text{loc},k+N_\text{p}-1} \\ \mathbf{0} & \cdots & \mathbf{0} \end{bmatrix}}_{\boldsymbol{\mathcal{E}}_\text{loc}} \underbrace{\begin{bmatrix} \boldsymbol{s}^{[\nu](\kappa)}_{\text{loc},k} \\ \boldsymbol{s}^{[\nu](\kappa)}_{\text{loc},k+1} \\ \vdots \\ \boldsymbol{s}^{[\nu](\kappa)}_{\text{loc},k+N_\text{p}-1} \end{bmatrix}}_{\boldsymbol{S}^{[\nu](\kappa)}_\text{loc}} \leq \underbrace{\begin{bmatrix} \boldsymbol{b}_{\text{loc},k} \\ \boldsymbol{b}_{\text{loc},k+1} \\ \vdots \\ \boldsymbol{b}_{\text{loc},k+N_\text{p}-1} \end{bmatrix}}_{\boldsymbol{B}_\text{loc}}. \tag{4.25}$$

And using the same approach as before the storage dynamics are integrated in with the inequality constraints.

$$\left(\boldsymbol{\mathcal{M}}_\text{loc}\boldsymbol{\Gamma} + \boldsymbol{\mathcal{E}}_\text{loc}\right) \boldsymbol{S}^{[\nu](\kappa)}_\text{loc} \leq \boldsymbol{B}_\text{loc} - \left(\boldsymbol{\mathcal{D}}_\text{loc} + \boldsymbol{\mathcal{M}}_\text{loc}\boldsymbol{\Phi}\right) \overline{SOC}^{[\nu]}_k \tag{4.26}$$

The local power balance is determined by solving the local constrained optimization problem

$$\begin{aligned}
\underset{\boldsymbol{s}^{[\nu](\kappa)}_{\text{loc},k}}{\text{minimize}} \quad & J^{[\nu](\kappa)}_{\text{loc},k} \\
\text{subject to} \quad & \\
& \boldsymbol{s}^{[\nu](\kappa)}_{\text{loc},k+l} \in \mathcal{S}^{[\nu]}_{\text{loc},k+l} \\
& \mathbf{2}^\text{T}_\nu \boldsymbol{s}^{(\kappa)}_{\text{glo},k+l} = \begin{pmatrix} (1 \;\; 1 \;\; 1 \;\; 1) & \mathbf{0} \\ \mathbf{0} & (1 \;\; 1 \;\; 1 \;\; 1) \end{pmatrix} \boldsymbol{s}^{[\nu](\kappa)}_{\text{loc},k+l} + \begin{pmatrix} \hat{p}^{[\nu]}_{\text{L},k+l} \\ \hat{q}^{[\nu]}_{\text{L},k+l} \end{pmatrix} \\
& \begin{pmatrix} \hat{p}^{[\nu]}_{\text{L},k} \\ \hat{q}^{[\nu]}_{\text{L},k} \end{pmatrix} = \begin{pmatrix} \overline{p}^{[\nu]}_{\text{L},k} \\ \overline{q}^{[\nu]}_{\text{L},k} \end{pmatrix} \\
& \left(\boldsymbol{\mathcal{M}}_\text{loc}\boldsymbol{\Gamma} + \boldsymbol{\mathcal{E}}_\text{loc}\right) \boldsymbol{S}^{[\nu](\kappa)}_\text{loc} \leq \boldsymbol{B}_\text{loc} - \left(\boldsymbol{\mathcal{D}}_\text{loc} + \boldsymbol{\mathcal{M}}_\text{loc}\boldsymbol{\Phi}\right) \overline{SOC}^{[\nu]}_k.
\end{aligned}$$

The goal of the optimization, is to find the best response to the given power profile by another controller (4.23) considering local costs and constraints. The constraints ensure that the power distribution is equal to the power required by the proposal of controller (κ) but local upper and lower bounds of controller $[\nu]$ are satisfied as well. Note that while in the local part of the cost function in **step 2** the total node power of subsystem ν proposed by subsystem κ is approximately weighted by C_kWh, in the local problem in **step 3** the nodal power $q^{[\nu]}_k$ is split according to (4.1) and each power is weighted differently. Consequently, after the reevaluation of the other proposals $J^{[\nu](\kappa)}_{\text{loc},k}$ must be sent to each controller (κ) in order to compute comparable cost in **step 4**.

Step 4: Comparison of the offset costs and implementation

After every controller has received the offset costs of the reevaluations, he computes the total cost for its own proposal which can be written as

$$J_{\text{tot},k}^{[\nu]} = J_{\text{glo},k}^{[\nu]} + J_{\text{loc},k}^{[\nu]} + \sum_{\kappa=1;\kappa\neq\nu}^{N} J_{\text{loc},k}^{[\kappa](\nu)}$$

and broadcasts the result. Every controller ν now evaluates the proposal with the lowest total cost

$$\nu^* = \text{argmin}_{\chi\in\mathcal{N}} J_{\text{tot},k}^{[\chi]}$$

and implements the corresponding local power distribution $\boldsymbol{s}_{\text{loc}}^{[\nu](\nu^*)}$. In the following an algorithm flowchart (Figure 4.2.3) is used to visualize the discussed four step negotiation algorithm.

4.2.3 Feedback of grid measurements for the approximated power flow update

In order to update the power flow approximation with the new operating point, apparent power and voltage magnitude values are needed. For both, measurement values are only available for the controlled part or node which is supervised by the controller ν. Values for other nodes have to be generated based on local calculations as pseudo-measurements. The power flow update process as depicted in Figure 2.8 is repeatedly started after each negotiation procedure to receive new matrices.

Real and pseudo apparent power measurements

After the implementation (**step 4**) each local controller knows the apparent power vector $\boldsymbol{s}_{\text{glo},k}^{[\nu](\nu^*)}$ and $\boldsymbol{s}_{\text{loc},k}^{[\nu](\nu^*)}$ from the proposal that has won the negotiation procedure. However, because he does not know the exact measurements of every node, the first value pair of apparent power is used instead to determined the update for the approximated power flow. This is reasonable, because these values represent set-points and as such, the implemented power by the inverters will be close to this value.

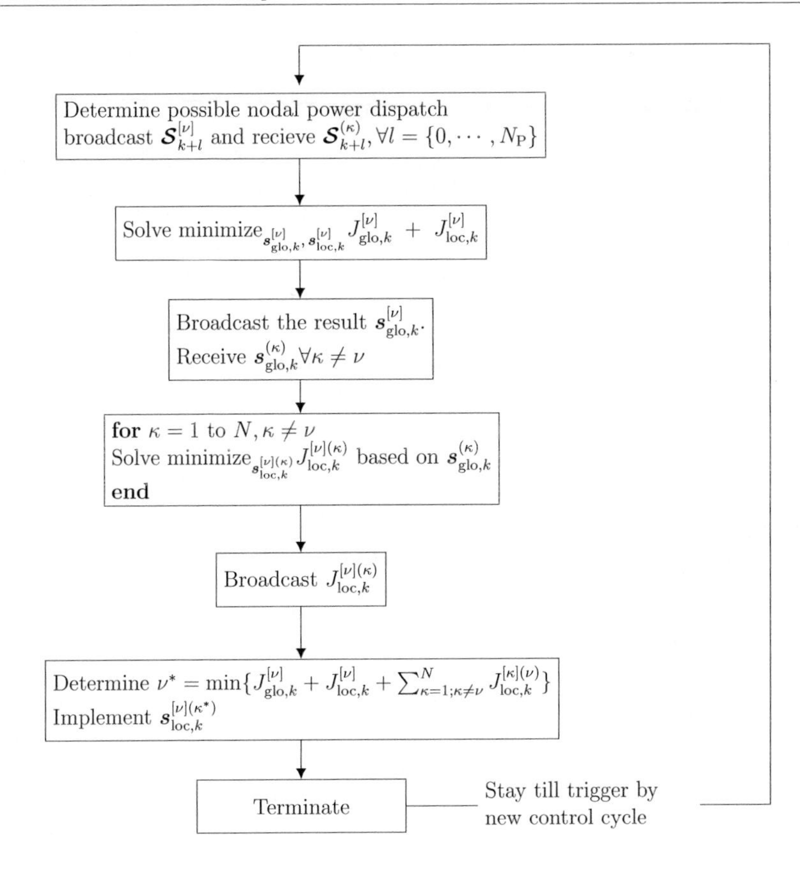

Figure 4.7: Algorithm flowchart depicting every single step of the four node negotiation in compact form

Real and pseudo voltage magnitude measurement

The local voltage magnitude from the smart-meter infrastructure in node ν can be used to parameterize the power flow approximation only for node ν. For the other nodes, the matrix $\boldsymbol{H}_{\mathrm{PF}}$ can be used together with $\boldsymbol{s}^{[\nu](\nu^*)}_{\mathrm{glo},k}$ to calculate an approximate nodal voltage for all other nodes κ.

4.3 Validation of the balanced distributed operation strategy

To show the effectiveness of the control approach, the algorithm developed in this chapter is applied to the four node and to the CIGRÉ benchmark grid shown in Figure 3.1 and Figure 3.2 respectively. The parameters of the lines and the transformer can be found in the appendix B and the power values of the used load profiles, PV and wind power units, as well battery storages, can be found from Table B.2 and Table B.1 respectively. In both benchmark test cases, it is assumed that a full controller coverage is possible, such that each node in the grid is supervised by a controller. The simulations for both cases are run over a full day (24 h) with different granularity of 1 min and 5 min prediction horizon of $N_{\mathrm{p}} = 5$ and $N_{\mathrm{p}} = 10$.

The validation is divided into two parts:

- First the performance of the algorithm with respect to the constraints is evaluated with the four node benchmark grid. As mentioned before, a safe operation close to the boundaries of the grid is desirable in order to maximize transport capacities. With regard to the BDEW traffic light system depicted in Figure 1.7, the algorithm performs in the green-phase. In this phase the market can minimize costs and maximize in-feed from distributed generation, without the intervention of a DSO. Electric vehicles are not included, to reduce complexity.
- The second part investigates how electric vehicles effect a grid with limited capacity, using the CIGRÉ residential low voltage grid as a benchmark. Especially, the charging behavior and the use of local battery storages is examined. Additionally, the tolerance of the distributed control approach to controller failure is presented in this setup, by disconnecting and reconnecting individual controller from the grid.

4.3.1 Operation at the power network boundaries

The four node benchmark grid is used to demonstrate the behavior of the distributed control algorithm under extreme situations. Enabling large consumers and renewable generation to be connected to the grid, even though the grid capability might not be initially constructed for it. Larger photo-voltaic plants can be installed, which increases the energy production in times with less solar radiance. A curtailment of active power is only pursued if the power system is at its boundaries and the battery storage can no longer receive charging power. In Figure 4.8 the simulation results of the distributed controller with regard to the nodal voltages of the grid are depicted. It is visible, that the grid is operated close to the upper $+10\%\,U_{\mathrm{Nom}}$ as well as to the lower $-10\%\,U_{\mathrm{Nom}}$

voltage boundary. The setup of the controllers for the individual nodes is summarized in Table 4.1. Values in Table 4.1 with the unit kWp describe a maximum value of the

Table 4.1: Simulation parameters four node benchmark with electric vehicles

Type	**Node ν**	**Value**
Photo-voltaic	$\{4\}$	300 kWp (3 ~)
Wind power	$\{2\}$	210 kWp (3 ~)
Hotel	$\{4\}$	60 kWp (3 ~)
Dairy farm	$\{2\}$	45 kWp (3 ~)
Household	$\{1,3\}$	10 kWp (3 ~)
Storage capacity $SOC^{[\nu]}$	$\{\text{all} \in \nu\}$	50 kWh
$SOC^{\text{min},[\nu]}$, $SOC^{\text{max},[\nu]}$	$\{\text{all} \in \nu\}$	$10\% SOC^{[\nu]}$, $90\% SOC^{[\nu]}$
Apparent power $S_{\text{c}}^{\text{max}}$, $S_{\text{d}}^{\text{max}}$	$\{\text{all} \in \nu\}$	22.5 kVA (3 ~)
$\eta_{\text{ch}}^{[\nu]}$, $\eta_{\text{d}}^{[\nu]}$	$\{\text{all} \in \nu\}$	90 %, 90 %
C_{Loss}, C_{kWh}		0.25 Euro
C_{DG}		−0.12 Euro
C_{ch}		0 Euro
C_{d}		0.15 Euro
N_{p}		10
T_{s}		1 min

load or in-feed profile. Furthermore, the 3~ sign resembles a three-phase value and 1~ a single-phase value.

Wind power plant at node two

At 2 a.m. the upper voltage boundary is reached at the wind power plant connected to node two. To increase possible active power in-feed to the grid, the inverter draws reactive power as depicted in Figure 4.10. Inductive power is increased up to the point where the current capacity of the cable is reached (Figure 4.9). The wind speed and as such the active power that could be provided to the grid is higher than the actual due to the additional reactive power support from the inverter. Thus the MPC has to reduce the active power and charge the local battery storage as shown in Figure 4.12. This way the surplus of power is channeled back to the grid at a time instant, where it is capable to receive it.

From the negotiation procedure results in Figure 4.13, it is possible to observe, that node two did not win the procedure in the time span between 2 a.m. to 4 a.m. in which the grid interaction took place. Instead node three won the negotiation and decided the dispatch solution of the second node. Such a scenario is possible because the power charts, that are exchanged prior to the negotiation include all the needed informations about the node.

Dairy farm at node four

The lower voltage boundary at node four is reached at 6 a.m. in the morning when the process of milking the cattle is at its maximum power point (Figure 4.8). Usually the high demand would violate the lower voltage constraint at this point, but instead, the controllers manage to operate the node close to the boundary. This is possible because the branch current between node three and four is below the nominal capacity of the cable (Figure 4.9). Capacitive reactive power can be provided to the grid by node three and four to raise the voltage. The controllers do not need to bring the battery storage into place, which would be the last resort to decrease the load power of the node.

Solar power plant at node three and four

During mid-day the solar radiation increases and the PV plants start to in-feed active power into the grid. The rated power of the PV plants is much higher than the benchmark grid would usually be able to integrate. As a result the voltage of node three and four are almost always at the upper boundary of the allowed range during midday, as depicted in Figure 4.8. To achieve a maximum of active power in-feed the nodes exchange a combination of inductive and capacitive reactive power with the grid (Figure 4.10). In Figure 4.9 the reason for the combination can be observed:

- The cable between node three and four is not at its current capacity limits. So node three compensates for some part of the reactive power.
- Node three only provides this type of support, because the cable between node three and the substation transformer is already at its limits. The line cannot handle additional current due to the inductive power injected from the node.
- This way, the two nodes maximize the in-feed of PV power from node four without compromising any grid limitations.

However, the support is not enough to make use of the full potential of photo-voltaic power at node four. The local battery storage is again used to shift the surplus of power to a later time instance, where it can be provided to the grid safely.

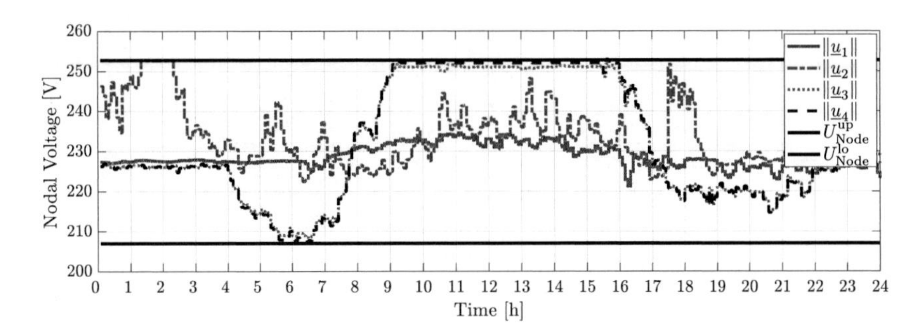

Figure 4.8: Grid operation close to the voltage constraints to maximize transport capacity

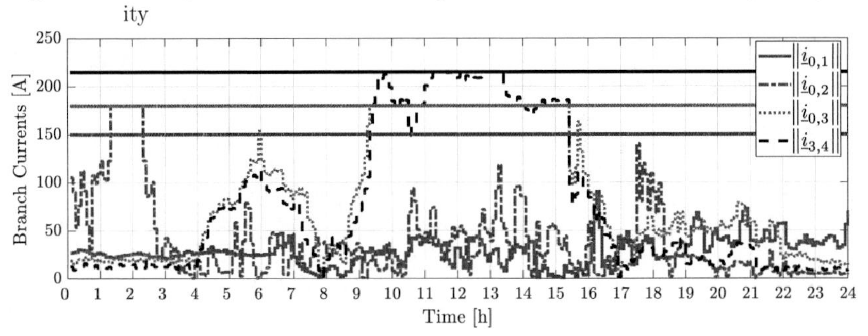

Figure 4.9: Grid operation close to the branch current constraints

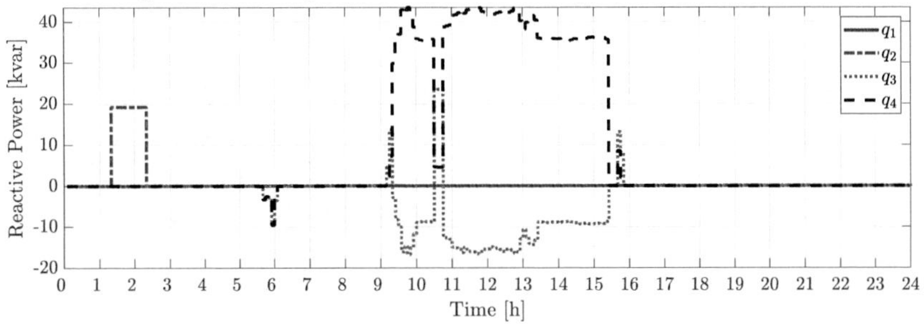

Figure 4.10: Use of flexibility in the form of reactive power in all four nodes of the benchmark grid

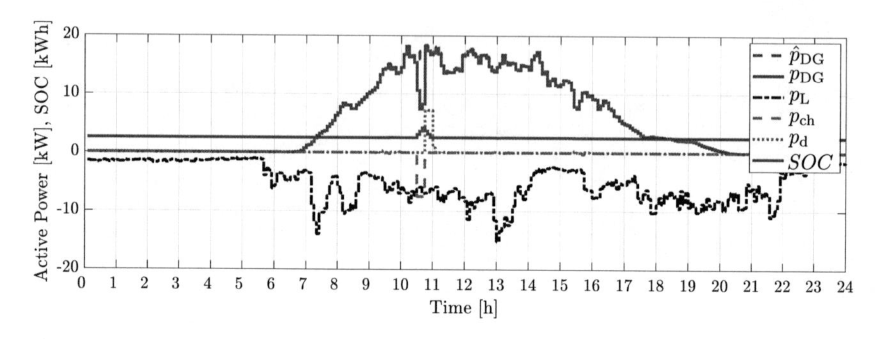

Figure 4.11: Local power distribution of node one

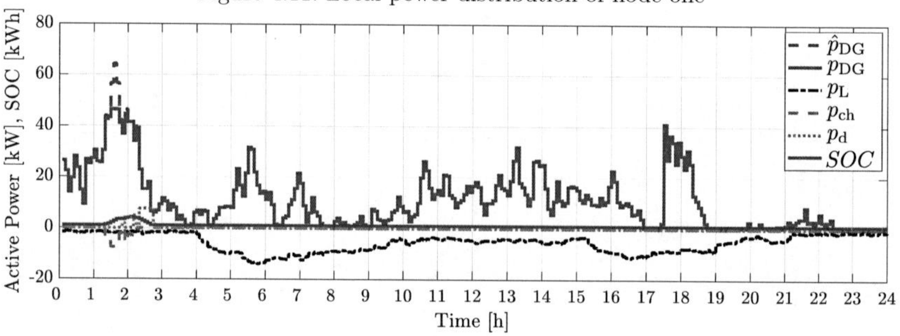

Figure 4.12: Local power distribution of node two

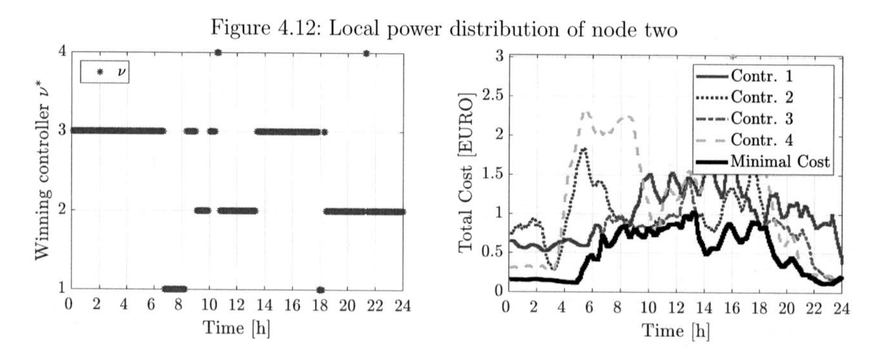

Figure 4.13: Comparision of winning controllers on the left and comparison of overall costs on the right

4.3.2 Operation with electric vehicles

Electric vehicles introduce additional flexibility to the grid if properly managed. In the control approach presented here, the charging produces a benefit for the consumer who is represented by a local cost function. Generic profiles are introduced that set the boundaries of possible power demand through charging of the electric vehicle. The local controller can adjust this power in the full range. Hence, the demand with regard to the grid operation is fully flexible. Furthermore, because this flexible power is not made globally available to other local controllers, the electric vehicle can only be charged if the controller wins the negotiation, or additional power from generation units is locally available. For the simulation study the grid is uniformly equipped with battery storages and photo-voltaic units. Moreover, at several nodes in the grid electric vehicles are positioned and connected to three-phase chargers. The parameters used for the simulation are summarized in Table 4.2. Photo-voltaic in-feed is assumed to be a clear sky approximation in this scenario. Thus the solar prediction and the actual power that is available are equal. Values in Table 4.2 with the unit kWp describing a maximum value of the load or in-feed profile.

Table 4.2: Simulation parameters CIGRÉ benchmark with electric vehicles

Type	**Node** ν	**Value**
Photo-voltaic	$\{\text{all} \in \nu\}$	30 kWp (3 ~)
Household	$\{\text{all} \in \nu\}$	5 kWp (1 ~)
Electric vehicles	$\{2, 5, 8, 10\}$	11 kWp(3 ~)
Storage capacity $SOC^{[\nu]}$	$\{\text{all} \in \nu\}$	20 kWh
$SOC^{\text{min},[\nu]}$, $SOC^{\text{max},[\nu]}$	$\{\text{all} \in \nu\}$	10%$SOC^{[\nu]}$, 90%$SOC^{[\nu]}$
Apparent power $S_{\text{c}}^{\text{max}}$, $S_{\text{d}}^{\text{max}}$	$\{\text{all} \in \nu\}$	12 kVA (3 ~)
$\eta_{\text{ch}}^{[\nu]}$, $\eta_{\text{d}}^{[\nu]}$	$\{\text{all} \in \nu\}$	90 %, 90 %
C_{Loss}, C_{kWh}		0.25 Euro
C_{DG}		−0.12 Euro
C_{EV}		0.25 Euro
C_{ch}		−0.15 Euro
C_{d}		0.15 Euro
N_{p}		5
T_{s}		5 min

Charging behavior for electric vehicles

The Controller behavior is depicted in the results of the simulation study, Figure 4.14 on the right hand side and Figure 4.15. Due to the large benefit, that the electric vehicle charging generates, controllers are able to win the overall cost comparison in their local charging slot. This can be observed for node eight between 4 a.m. and 8 a.m. as depicted in Figure 4.14 on the left. Some time period is thus available for the controllers to fill the car batteries. A second scenario can allow for the local charging – the global solution of the negation is to in-feed less power than locally generated. The additional active power is thus fed into the electric vehicle. This behavior can be observed in Figure 4.15 for node eight between 8:15 a.m. and 12:15 p.m. during the day.

The charging profile depicted in the dashed red line, can be interpreted as a window, given by the owner of the car. While the calculated red solid trajectory can serve a reference for the charging controller of the car. Overall costs for the grid operation and additional costs for each node are always considered during the four-step negotiation procedure between the controllers. Hence, the combination of calculated charging power and time instances are the best option available.

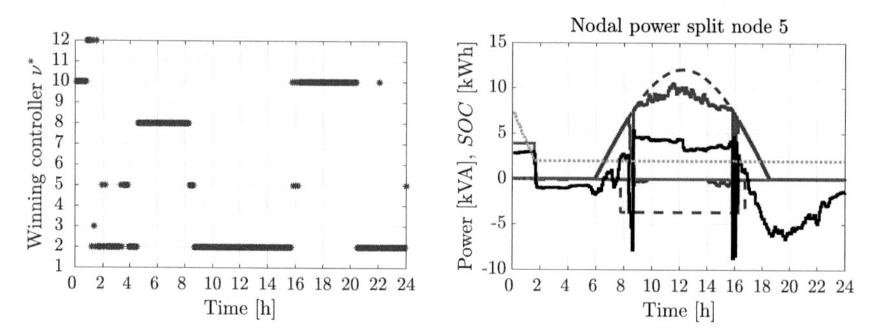

Figure 4.14: Simulation result with the winning controller on the left, and the local power distribution of node two on the right

Investigation of battery usage

In many cases, the battery state of charge is scheduled by an upper control layer, that has some form of predictor for load and in-feed. For such a static planning of the storage

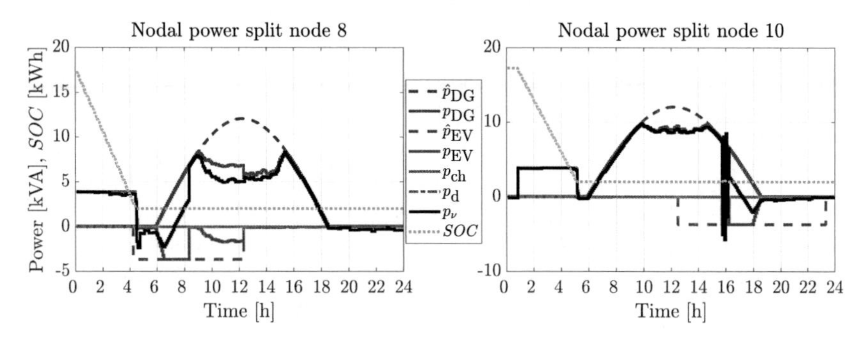

Figure 4.15: Simulation results of the local power distribution of node eight and node ten

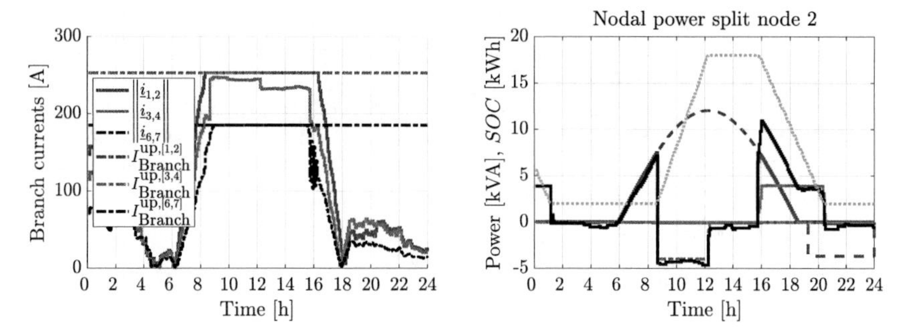

Figure 4.16: Usage of the battery storage in node two, to mitigate an over-current in the main branch of the benchmark grid

usage in the grid, power predictions must be very accurate. In general this assumption is not valid, as has been already pointed out before. Instead, here, the battery is directly controlled without using an external reference value, to react to possible disturbances during operation. This point is especially important in case the controllers disconnect from the grid.

The battery storage, which is connected to each of the nodes, can be used to store energy from the surplus of power generation. This way, power that would have to be curtailed, is stored locally. In the simulation results, depicted in Figure 4.15, this cannot be observed as a general behavior, because system limits are only reached during maximum in-feed times of the photo-voltaic units. Only the controller located at node two, uses the battery storage to its full extend, to mitigate the effect of an over current

in the main branch, as depicted in Figure 4.16.
It should be noted, that only the branch currents that reach the cable ampacity rating are depicted for a better overview.

Not only is the battery charged and discharged, the local production is also reduced to zero between 10 a.m. and 2 p.m. to keep the branch current inside its physical limits. From the negotiation results in Figure 4.15 it is visible, that the behavior is not inflicted by the proposal of other controllers, but is supported by controller number two.

4.3.3 Controller failure simulation

The resilience of the operation strategy against failure of one or several controllers is one of the main motivations for the use of distributed control schemes. An operation of the grid must be possible, even though controllers go off-line. In the following simulation example two controllers are detached from the grid and subsequently the negotiation procedure, at different time instances during the day. The scenario analyses the behavior of the overall operation and the uncontrolled node. The setup for this simulation, summarized in Table 4.2 is again used. The controllers are detached and reconnected to the grid at the following time instances:

- Controller at node two is off-line for several hours from 10 a.m. until 4:20 p.m.
- At node eight the controller is deactivated between 8.30 a.m until 9:30 a.m.

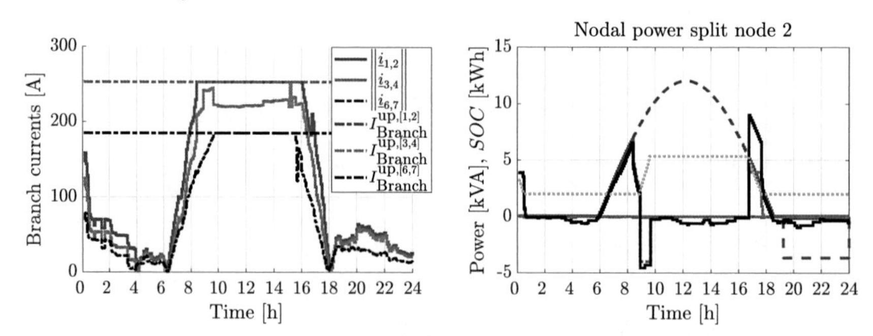

Figure 4.17: Simulation results of the branch current magnitudes on the left and controller failure in node two on the right

From the results it can be observed that the voltage profile and the branch current limitations are still maintained, as depicted in Figure 4.18 and Figure 4.17 respectively. This

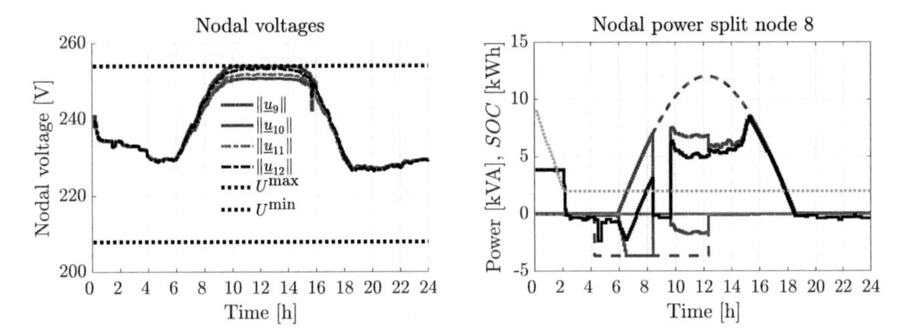

Figure 4.18: Simulation results of the node voltage magnitudes on the left and controller failure in node eight on the right

is possible, because every controller has global information about the power network. An operation of the grid with the four step negotiation is still possible for the remaining controllers and system can be safely operated.

However, load predictions, included in the constructed power charts, are not available anymore for the nodes with controller failure. Instead measurement values from the smart-meter infrastructure must be used to monitor the load demand of uncontrolled nodes. Thus, only a single value rather than a prediction over the full controller horizon is available.

4.3.4 Intermediate conclusion

The controller behavior shows that the usage of the local battery storage is only favored if system constraints must be kept. Otherwise it's use is too expensive and does not make sense from a local perspective. Production of renewable energy is maximized if possible and reactive power is exchanged with the grid whenever needed. However, the reactive power support is kept minimal to reduce ohmic losses in the power lines. Furthermore, the controllers effectively use their charging slot for electric vehicles without the need to broadcast this particular flexibility to others.

The investigation of possible controller failure shows that the approach is resilient against single or multiple controller outages. A detachment from the distributed control structure and the grid is easily possible with the approach and does not counteract the efficient operation. In summary, the following results have been achieved:

- System limitations are maintained under every circumstance and load situation.

- The four step negotiation procedure has a fair distribution of winners, over all possible nodes and controllers.
- Charging electric vehicles at the nodes is possible, even though controllers do not share information about this flexible demand.
- Controller failure can be efficiently handled and a safe operation of the grid is still possible.

4.4 Practical verification of the balanced distributed operation strategy

The controller setup, which was described in this section, is now implemented on a demonstrator platform. This platform consists of two rapid prototyping systems (RPS). The first one is used for the controller and the second one emulates the 40 node LV grid as has been described in the benchmark system setup depicted in Figure 3.3 from Section 3. Both systems are connected via a data interface and interact with a three-phase grid hardware model, which is a scaled medium voltage grid (see Figure 3.4 for a schematic representation). Furthermore, all parameter values of the grid cables and transformer can be found in the appendix B in Table B.6. Values with regard to the power ratings of each node can be found in Table 4.3.

4.4.1 Multiprocessor controller platform

The distributed control strategy is implemented on a dSpace© DS1006© multiprocessor platform with four central processing units (CPUs) (see Figure 3.4). The controllers are distributed onto the physical cores of the rapid prototyping system (RPS) and run in parallel as individual local controllers.
These individual cores of the RPS are configured in a multiprocessor (MP) setup and communicate over the internal IPC bus of the dSpace© framework. This communication link is necessary to exchange the performance charts and other relevant information of each controller used for the operation strategy (see Figure 4.4).
The optimization problem is generated and solved on-line in every processing unit in a fully distributed fashion. This way the algorithm is embedded in a close to reality setup. In order to solve the optimization problem, which is a quadratically constrained quadratic program (QPQC), an Interior-Point method that can be compiled for the RPS based on the Mehrotra predictor-corrector method was developed. It is not further introduced in this thesis, but can be found in [Dom13, Ken17].
The distributed control strategy is executed with a prediction horizon $N_{\mathrm{p}} = 5$ and

a sample-time $T_s = 1\,\text{min}$. In the real world time scale the individual optimization problems running on the cores of the RPS only need 500 ms-750 ms. The configured control cycle time of the dSpace is consequently set to 1 s.

4.4.2 Low voltage grid emulation and aggregation

The 40 node low voltage grid, as presented in Figure 3.3 is emulated on the second RPS which is coupled to the controller RPS over a bidirectional data interface. Over this data interface, measurement values from the emulated grid as well as the set-points for inverters are transmitted. For the emulation an FBS method is executed to solve the power flow in real-time. Domestic demand, dairy farm and hotel load as well as PV profiles are stored in a database on the RPS with a time granularity of 1 min.
To account for the real-time feedback of the grid the data is scaled up using a linear interpolation method. The sample time of 20 ms is chosen due to the fact, that the grid frequency is ≊50 Hz and only after one period of the voltage a real active power value can be calculated [AWA17]. Additionally, white noise is added to the signals to account for possible measurements errors.

For the verification in the demonstrator it is assumed, the that the network is only

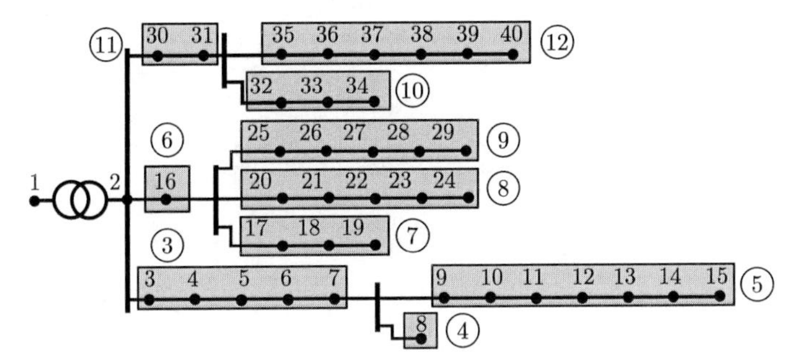

Figure 4.19: Original 40 node and aggregated 12 node low voltage grid in grey boxes ©IEEE2018

equipped with four controllers. Nodes that are not supervised by a controller, still have smart-meters, smart-meter-gateways and a communication interfaces available. To realize this setup, the 40 node benchmark grid is aggregated. Only nodes with a high impact on the nodal voltages and branch currents are chosen and grouped as controlled areas. These nodes are households, hotels or farms with additional electric vehicles and distributed generation. Other nodes are concentrated by simply adding up their respective branch impedances and their measured demand.

The original node numbering is highlighted in the gray boxes and the resulting aggregated and grouped node numbering is displayed inside of white colored circles as depicted in Figure 4.19. The approach results in a reduced representation of the power grid that characterizes the main behavior. The nodal setup of the aggregated power grid can be found in Table 4.3 as Agg. Node μ. Areas that will be supervised by a controller for the implementation are 4, 5, 9, 12.

Evaluation of the aggregation quality

The quality of this system reduction is important for the success of the control method. Hence, the power flow for the grid is evaluated for both the full as well as the reduced representation. The simulation result of this comparison is depicted in Figure 4.20

Table 4.3: Node parameters for the verification using the 40 node benchmark grid

Distributed generation	**Node ν/Agg. node μ**	**Max. value**
Photo-voltaic	$\{16,31\}/\{6,11\}$ $\{15,19\}/\{5,7\}$ $\{24,7\}/\{8,3\}$ $\{34\}/\{11\}$ $\{8,40\}/\{4,12\}$	24 kWp (3 ~) 24 kWp (3 ~) 32 kWp (3 ~) 150 kWp (3 ~) 285 kWp (3 ~)
Load demand	**Node ν/Agg. node μ**	**Max. value**
Dairy farm	$\{8,40\}/\{4,12\}$	40 kWp (3 ~)
Hotel	$\{15,34\}/\{5,11\}$	60 kWp (3 ~)
Households	Rest	5 kWp (3 ~)
Electric vehicles	$\{15\}/\{5\}$ $\{8,19\}/\{4,7\}$ $\{34,40\}/\{11,12\}$	$4\times$ 22 kWp (3 ~) 22 kWp (3 ~) 22 kWp (3 ~)
Stationary battery	**Agg. node μ**	**Value**
Storage capacity $SOC^{[\nu]}$	$\{4,12\}$ $\{5,9\}$	60 kWh (3 ~) 30 kWh (3 ~)
$SOC^{\mathrm{min},[\nu]}$, $SOC^{\mathrm{max},[\nu]}$	$\{4,5,9,12\}$	$10\% SOC^{[\nu]}$, $90\% SOC^{[\nu]}$
$P_{\mathrm{c}}^{\mathrm{max},[\nu]}$, $P_{\mathrm{d}}^{\mathrm{max},[\nu]}$	$\{4,5,9,12\}$	22,5 kW (3 ~)
$\eta_{\mathrm{c}}^{[\nu]}$, $\eta_{\mathrm{d}}^{[\nu]}$	{all}	90 %, 90 %

The dotted black line in sub-figure a) depicts node eight, sub-figure b) node 15 of the voltage magnitude for the original 40 node low voltage grid. The solid black lines repre-

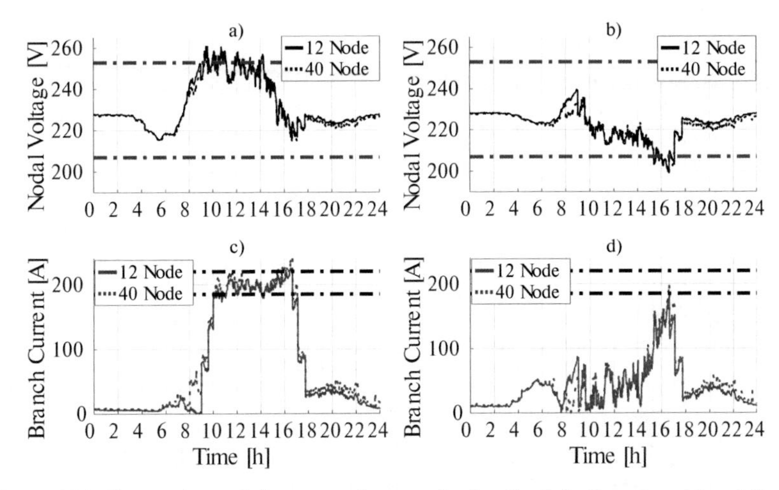

Figure 4.20: Comparison of the power flow results for the full 40 node grid and its 12 node aggregated representation

sent the power flow results using the aggregated grid. In sub-figure c) the branch currents between node 9-15 are evaluated and in sub-figure d) the ones for node 3-7. Both simulations only show a small difference in the power flow results for branch current and nodal voltage magnitudes. Both electric vehicle chargers and distributed generation are considered in the aggregated grid, such that the significant influences are covered by the reduced representation.

4.4.3 Estimation of the low- and medium voltage grid interaction

For the grid evaluation, it is advantageous to not only calculate the power flow for the low voltage grid side, but also to know the interaction between low voltage grid emulated on the RPS and medium voltage grid represented by the hardware grid. This way, the rather strong assumption of a slack bus at the point interconnecting the grids can be relaxed. The balanced three-phase voltage and current of the virtual substation transformer is measured as depicted in Figure 4.21. At the same time a power flow is solved simultaneously for the 20 kV grid depicted in Figure 3.5 to evaluate the estimation quality. All relevant input data for his power flow calculation is monitored by measurements devices and processed by the controller RPS on a low level. The resulting virtual measurements can finally be included in the power flow approximation for the

40 node low voltage grid substituting $\underline{u}_0$ and $\underline{\boldsymbol{E}}_0$ from Equation 2.23 with a measured value $\overline{\boldsymbol{u}}_0$ and $\overline{\boldsymbol{E}}_0$ respectively. Even though in a real setup there will be certain high frequent fluctuation in the voltage, it is assumed that during the time intervals of the control cycle (1 sec.), the voltage mean value at the substation transformer does not change significantly.

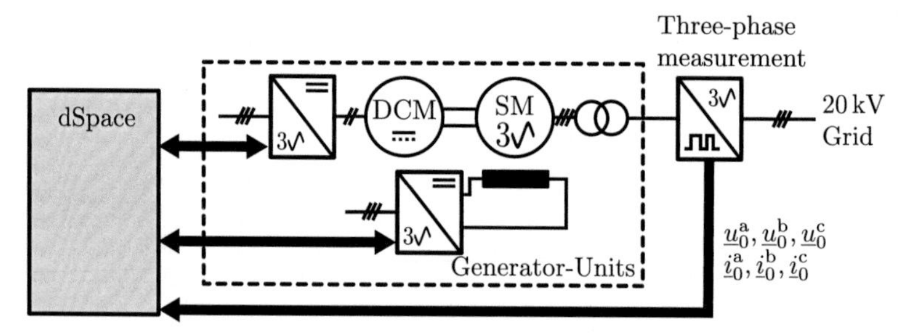

Figure 4.21: Experimental setup for the voltage estimation at the virtual substation transformer

4.4.4 Implementation results with the rapid-prototyping-system

The following subsection present the result from the implementation of the distributed control algorithm in combination with the smart-grid demonstrator described in Section 3. Large amounts of electric vehicles are integrated in the grid, to evaluate their effect on the constraint satisfaction of the strategy. Moreover, the prediction of renewable generation is now combined with real measurement values from the field. Controllers use the prediction to calculate a possible dispatch. The actual real measurement values are then injected by the local inverters, if they are smaller than the prediction, else they directly use the predicted values.

Result: grid operation with the distributed controllers

When several electric vehicles arrive at node 15 as depicted in Figure 4.22 right lower subfigure, the MPC limits the maximum (red dashed line) charging power to the resulting black solid line. The reason for this is twofold: first around midday to account for the branch line (11-12) (magenta solid line) thermal constraints of 185 A (black dashed line)

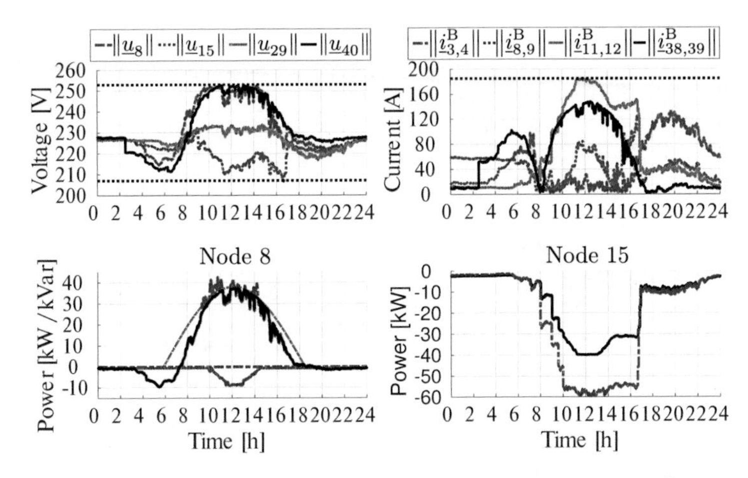

Figure 4.22: DMPC implementation result of the four controlled nodes ©IEEE2018

as depicted in Figure 4.22 right upper sub-figure. Secondly, when the in-feed of PV plants in the grid decreases between 4-6 p.m., the high EV charging power would violate the voltage constraints of 207 V (dashed black line) of node 15 (dotted blue line) as shown Figure 4.22 left upper sub-figure. The limited charging power is now no longer branch current but nodal voltage related. At the same time in midday the in-feed of the PV plant at node 8 (Figure 4.22 left lower sub-figure, black solid line) is maximized to the predicted value (dashed magenta line). It should be noted that the in-feed prediction integrated in the distributed MPC does not include the effect of rapid cloud coverage changes and thus estimates a mean power value. The DMPC uses the prediction as an upper bound, resulting in the solid black line in Figure 4.22 left lower sub-figure being below the red solid one. Additionally, the PV inverter of node 8 draws reactive power from the grid to reduce the voltage rise (Figure 4.22 left lower sub-figure, blue solid line). But does this only when it it necessary, because it increases the grid losses.

Result: estimation of the medium voltage grid interaction

For the medium voltage side estimation of the 20/0.4 kV substation transformer the LIPF is calculated for the general ZIP-bus model (2.27). This has to be done because now both shunt and load impedances are present in the circuit. For the power flow calculation on the demonstrator side, the voltage of the distribution transformer on the right hand side depicted in Figure 4.21 is now used as the slack bus for the power flow calculation.

The results in Figure 4.23 show that the high voltage side of the substation transformer is changing rapidly, but that the LIPF achieves a good estimation of the medium voltage grid interconnection (solid line) compared to a directly measured value (dashed black line). This can be said for both the current (red) and voltage magnitude (blue) estimation. Only the positive-sequence of both voltage and current are used for the comparison.

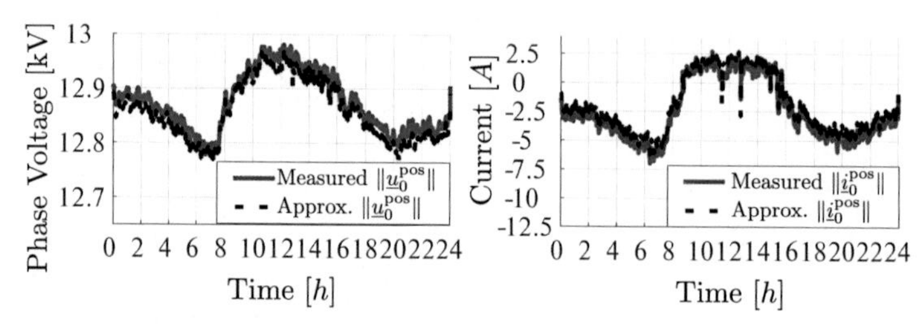

Figure 4.23: Measured and estimated substation transformer voltage and current ©IEEE2018

The impedances change of the transmission lines due to heating effects have not been characterized individually for the hardware grid, so that the differences in voltage and current can be caused by parameter uncertainties.

4.4.5 Intermediate conclusion

The experimental verification on a dSpace© rapid prototyping system shows that a real-time implementation of the distributed control approach is feasible. With a process execution time of 1 s there is enough time for communication between the local MPCs in the different nodes in the grid regarding a real world control cycle of several minutes. Additionally, by estimating the high voltage side at the 20/0.4 kV substation of the controlled grid the assumption of a slack bus at the root node could be relaxed. The approximation used in the controller is fed with the transformed value of the estimated voltage.

5 Utilizing flexible resources for unbalance mitigation in low voltage grids

European low voltage grids are subject to unbalance in nodal voltages and currents. This mainly originates from the fact that household consumers and smaller photo-voltaic units are only connected to a single phase of the distribution grid. The following section introduces a distributed optimization algorithm that utilizes the power flow approximation developed in Section 2.4.1 to mitigate the unbalance by using specific properties of the connected generation units, battery storage, and electric vehicle chargers. By transforming the power flow approximation into symmetrical-components it is possible to mitigate the sequence-components, representing the unbalance, in a selective or unified manner. This section was partly presented in [BBL17]

5.1 Introduction and related work

Usually, it is desirable to distribute the consumers equally onto the three phases throughout the low voltage grid [Sho04]. In accordance to DIN-EN-50160 the voltage unbalance caused by single-phase consumers and generation is restricted to a value of $\leq$2%. This limitation is motivated by the significant impact the voltage unbalance has on the grid infrastructure [AMS13]. If for example one of the phases is heavily loaded, the coupling can have an undesired effect on the other phases e.g. a voltage rise. Another direct effect is an increase in losses in transformers and electrical machines [FSZ10].
Unbalance mitigation for power grids has been investigated by using phase switching [CC00, SWG14], Scott-T transformers [SJK10, AMK12, LC13], as well local inverter balancing [CLT^{+}12][SMW14]. Phase switching approaches change the order of the three phases to move the loading from one phase to the other. This way a certain re-balancing can be achieved. Scott-T transformers can be used to supply two single-phase loads from a three-phase voltage. Through constructional features, the power drawn from the three-phase grid is fairly balanced. However, these two groups involve costly grid equipment like automatic power switches or custom-made transformers to be installed in the grid. The third group involves the use of power electronic equipment which is already

installed in the grid, because it is part of a photo-voltaic plant or battery storage. Most algorithms in this group use a decentralized control approach and have no knowledge of the underlying electrical network. Even though the voltage balancing can be solved locally to some extent, there are cases, where the unbalance is increased if there is no grid knowledge or coordination between the controllers available [Ben15]. Furthermore, the negative- and especially the zero-sequence component cannot be estimated precisely without knowing the frame it refers to. Hence, methods which use a model of the grid and solve a power flow formulation, deliver a more exact result, as they can calculate every voltage phase and magnitude. If the power system is unbalanced, all three phases of the voltage can have a different magnitude and phase. Furthermore, the neutral conductor transports current in case the load is unbalanced [Ker12]. A three-phase four-wire representation of the system is necessary [CFO03].

Efficient three-phase power flow algorithms to solve problems in unbalanced distribution system have been intensively studied in the literature, see [ZC95],[CS95],[ER99],[AANR05] and the references therein. Unfortunately, only a limited amount of scenarios can be found which apply a three-phase optimal power flow (TOPF) to solve unbalance related issues like mitigation or the reactive power-sharing problem [BLR+11, APCP13]. This is mainly due to the high computational effort to solve a full alternating current (AC) version of the TOPF.

In this regard centralized solutions of the TOPF with the help of power flow approximations and relaxations have been developed [DGW12, BHK+15, ASDB16]. They are computationally much more efficient than their non-linear AC counterparts, but still, need to gather system information at one central point. Algorithms that solve the TOPF in a distributed fashion further reduce the computational burden by decomposing it into sub-problems which are less demanding to solve. Especially algorithms based on the alternating method of multipliers (ADMM) have received a lot of attention in recent years, see [GL14, AZG13], as well as a comprehensive review in [MDS+17].
The use of the TOPF for unbalance mitigation can be grouped into two possible directions:

- Tracking of a nominal reference in the form of a balanced voltage [ASDB16, KZGB16, Ayi17, RKK19]
- Minimizing some defined unbalance criterion [APCP13, Ben15, CCM+18]

The tracking of a voltage reference can again limit the operational flexibility of an approach. Hence, to maximize the transport capabilities of the low voltage grid, it is not always desirable to operate the system close to the nominal voltage. Methods that minimize some predefined unbalance criteria can achieve a balancing of the grid without using a fixed reference. However, for this type of approach, a more detailed power flow approximation is necessary. It needs to output both magnitude and phase information because otherwise, it does not estimate correct unbalance factors. The LinDistFlow model, which is widely used for efficient implementations of optimal grid balancing approaches, cannot be utilized for this purpose. It only outputs magnitude information which is not sufficient to estimate the negative- and zero-sequence component of the voltage.

To the authors best knowledge distributed algorithms that solve three-phase optimal power flow with a proper system representation have only been proposed with semidefinite programming (SDP) relaxations [GL14] and the use of ADMM [PL15] together with the non-linear DistFlow model. These approaches would also be capable of implementing an unbalance criterion minimization, even though the before mentioned authors did not follow this direction.

5.1.1 Setup of the distributed controllers to mitigate voltage unbalance

Proposed solution approach

To achieve the balancing, an explicit local representation of the unbalance in the grid based on the introduced power flow approximation in Section 2 is proposed. The approximation is capable of a detailed representation of negative- and zero-sequence components [SSM07], with only smart-meter measurements available in the different nodes of the grid. Because the model is linear, it can be decomposed, so that the local power injections in the individual nodes can be mapped on their system-wide effect on the three-phase grid voltages. The mapping linearly relates apparent power in the nodes to the negative- and zero-sequence voltage component of every other node in the grid. Every node thus knows how it can influence the unbalance in the whole power network.

A strategy based on distributed optimization with a Jacobi type algorithm is developed, that is suitable for optimization problems that have a separable cost function [NND11]. The method does not need a central coordinator and is feasible in every iteration, such that even though the balancing strategy has not yet converged, a communication failure does not render it infeasible. The objective of the local optimization problems is to directly reduce the sequence-components instead of tracking the nominal voltage as a reference [ZDGD13, KZGB16]. With these approaches, the balancing controller would tie the phase voltages to the nominal value, which would counteract the upper layer MPC that maximizes the transport capacity of the grid. Furthermore, this way, the balancing is more flexible, as the decomposed representation of negative- and zero-sequence components can be minimized separately, or unified. Three-phase, as well as single-phase units, are integrated and commit their flexibility for the mitigation strategy. Especially the integration of controllable single-phase electric vehicle chargers is important, as they significantly contribute to the unbalance in low voltage grids [SGLZ11].

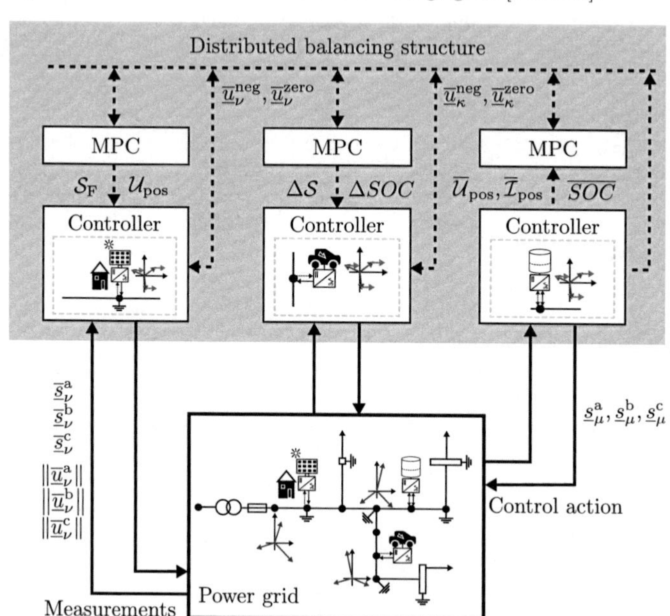

Figure 5.1: Distributed control structure for the unbalance mitigation in low voltage grids presented in this work

Interaction between balanced operation and mitigation strategy

The balanced operating strategy from Chapter 4 takes care of the systems-wide constraints on the assets of the grid. This means that the positive-sequence voltage and branch current are always kept in certain predefined bounds and an underlying unbalance mitigation strategy only needs to take care of local constraints like inverters capabilities. Moreover, it is assumed that the balancing approach presented here, is coupled with the MPC in every node, in order to receive calculated set-points for fixed balanced apparent power injections $\mathcal{S}_\mathrm{F}$ and positive-sequence voltage $\mathcal{U}_\mathrm{pos}$ to parameterize the three-phase power flow approximation. While from three-phase measurements, the local positive-sequence voltage and current $\overline{\mathcal{U}}_\mathrm{pos}$, $\overline{\mathcal{I}}_\mathrm{pos}$ is fed back to the MPC to parameterize the balanced single-phase power flow approximation. Furthermore, the balancing controller receives information about the flexibility that three-phase inverters $\Delta\mathcal{S}$, and the battery storage ΔSOC can provide to support the mitigation strategy, as depicted in Figure 5.1. For this purpose the current measured state of charge of the battery is returned to the MPC in form of $\overline{SOC}$. Single-phase inverters, on the other hand, can be fully used for the strategy, as they are not considered in the upper layer MPC for balanced operation. The three-phase quantities are further denoted as calligraphic variables of the following form

$$\mathcal{S}_k^{[\nu]} = \left(\left(\mathcal{P}_k^{[\nu]}\right)^{\mathrm{T}} \left(\mathcal{Q}_k^{[\nu]}\right)^{\mathrm{T}}\right)^{\mathrm{T}}, \mathcal{P}_k^{[\nu]} = \left(p_k^{\mathrm{a},[\nu]} \quad p_k^{\mathrm{b},[\nu]} \quad p_k^{\mathrm{c},[\nu]}\right)^{\mathrm{T}}, \mathcal{Q}_k^{[\nu]} = \left(q_k^{\mathrm{a},[\nu]} \quad q_k^{\mathrm{b},[\nu]} \quad q_k^{\mathrm{c},[\nu]}\right)^{\mathrm{T}}. \tag{5.1}$$

Balanced powers from the MPC are translated into fixed values for the local balancing controllers to determine flexibility reserves.

5.1.2 Preliminaries to the balancing approach

Grid unbalance indicators

To assess the unbalance in a grid based on the sequence-voltage of a node ν, the local negative-sequence (5.2a) and zero-sequence unbalance factors (5.2b) according to DIN-EN-50160 [DE11] are evaluated as following

$$\lambda_\nu^{\mathrm{neg}} = \frac{||\underline{u}_\nu^{\mathrm{neg}}||}{||\underline{u}_\nu^{\mathrm{pos}}||} \tag{5.2a}$$

$$\lambda_\nu^{\mathrm{zero}} = \frac{||\underline{u}_\nu^{\mathrm{zero}}||}{||\underline{u}_\nu^{\mathrm{pos}}||}. \tag{5.2b}$$

There are several other unbalance indicators that can be found in the literature [SSM07]. However, (5.2a) and (5.2b) are seen as the most exact and the same time most difficult to acquire [GMR19]. Most approaches in literature investigate solely the negative-sequence components (5.2a). This is only sufficient for medium- or high voltage grids due to their consumer structure. In the case of low voltage feeders the zero-sequence component plays a significant role, as demonstrated by benchmark results with OpenDSS [Dug10] for the IEEE European low voltage Test-feeder [PES15].

Approximated unbalance indicators and their decomposition

The approximated power flow developed in Sec. 2.4.2 can now be used to represent the phase- and sequence-frame quantities of the grid. With the decomposition (2.33) the unbalance indicators (5.2a) and (5.2b) can be approximately expressed as

$$\lambda_\nu^{\text{neg}} \approx \frac{\left\|\mathbf{2}_\nu^{\text{T}} \mathbb{M}_{\text{neg}} \Delta \boldsymbol{\mathcal{U}}_{\text{pnz}}\right\|}{\left\|\mathbf{2}_\nu^{\text{T}} \mathbb{M}_{\text{pos}} \boldsymbol{\mathcal{U}}_{\text{pnz}}\right\|} \tag{5.3a}$$

$$\lambda_\nu^{\text{zero}} \approx \frac{\left\|\mathbf{2}_\nu^{\text{T}} \mathbb{M}_{\text{zero}} \Delta \boldsymbol{\mathcal{U}}_{\text{pnz}}\right\|}{\left\|\mathbf{2}_\nu^{\text{T}} \mathbb{M}_{\text{pos}} \boldsymbol{\mathcal{U}}_{\text{pnz}}\right\|}. \tag{5.3b}$$

Both negative- and zero-sequence of each node can be determined as a relative component from $\Delta \boldsymbol{\mathcal{U}}_{\text{pnz}}$ by a linear mapping through $\mathbb{M}_{\text{neg}}$ and $\mathbb{M}_{\text{zero}}$ from (2.34).

To demonstrate this decomposition in an example, the following illustration (Figure 5.2) is introduced. It shows the relationship of the decomposed sequence components for node $\nu = 1$ and phase {a,b,c}. The grey arrows $\underline{u}_0^{\text{a}}, \underline{u}_0^{\text{b}}, \underline{u}_0^{\text{c}}$ represent the balanced slack bus voltage and the black solid line arrows $\underline{u}_1^{\text{a}}, \underline{u}_1^{\text{b}}, \underline{u}_1^{\text{c}}$ the actual phase voltage. The colored arrows depict how the individual sequence components $\underline{u}_1^{\text{pos}}, \Delta\underline{u}_1^{\text{neg}}, \Delta\underline{u}_1^{\text{zero}}$ add up to represent the phase voltage. The individual components of Figure 5.2 can be

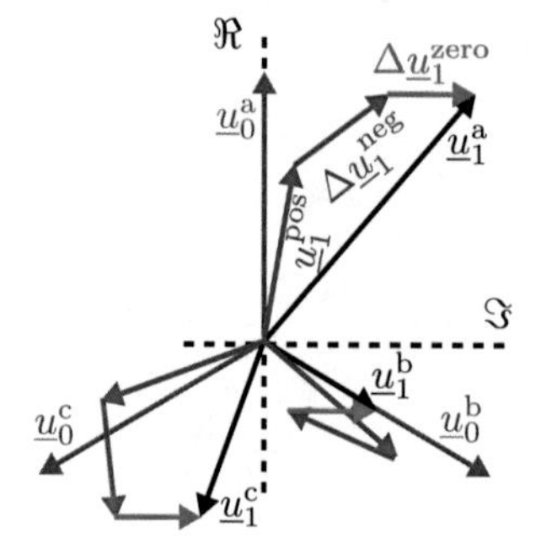

Figure 5.2: Exemplary symmetrical sequence-decomposition of the unbalanced phase voltages of node 1

calculated using the following formulation

$$[\Re\{\Delta\underline{u}_1^{\text{pos}}\}, \Im\{\Delta\underline{u}_1^{\text{pos}}\}]^{\text{T}} = \mathbf{2}_1^{\text{T}} \mathbb{M}_{\text{pos}} \Delta \boldsymbol{\mathcal{U}}_{\text{pnz}} \tag{5.4a}$$

$$[\Re\{\underline{u}_1^{\text{pos}}\}, \Im\{\underline{u}_1^{\text{pos}}\}]^{\text{T}} = \mathbf{2}_1^{\text{T}} \mathbb{M}_{\text{pos}} \Delta \boldsymbol{\mathcal{U}}_{\text{pnz}} + \boldsymbol{\mathcal{U}}_0 \tag{5.4b}$$

$$[\Re\{\Delta\underline{u}_1^{\text{neg}}\}, \Im\{\Delta\underline{u}_1^{\text{neg}}\}]^{\text{T}} = \mathbf{2}_1^{\text{T}} \mathbb{M}_{\text{neg}} \Delta \boldsymbol{\mathcal{U}}_{\text{pnz}} \tag{5.4c}$$

$$[\Re\{\Delta\underline{u}_1^{\text{zero}}\}, \Im\{\Delta\underline{u}_1^{\text{zero}}\}]^{\text{T}} = \mathbf{2}_1^{\text{T}} \mathbb{M}_{\text{zero}} \Delta \boldsymbol{\mathcal{U}}_{\text{pnz}} \tag{5.4d}$$

representing the positive-, negative- and zero-sequence component respectively. Note that the components can only be applied to phase **a** without a rotation. For the other phases the inverse of $\underline{\boldsymbol{T}}_{\text{pnz}}$ can be used to map the components respectively [Das17].

Structural properties of radial grids and control area sectioning

The admittance matrix (2.1) of radial distribution grids has a sparse structure. It's counterpart the impedance matrix is dense, but still has some distinct structural properties. The structure reflects the inter-coupling of different nodes and consequently the phases in the grid [KZGB16]. For some control purposes this can mean that it is not necessary or even counter-productive to consider the grid as a whole. Especially, if the grid is built as depicted in Figure 5.3 on the left, where the inter-coupling of the three main branches is rather weak as can be seen on the right part. They are only coupled in the short circuit impedance of the substation transformer. Branch impedances on the other hand, have another order of magnitude in value and thus dominate the matrix. That is why the balancing between two respective main branches is not efficient as they are loosely coupled and require high amounts of local power injection to influence each other. The grid is sectioned by separating it into the main branches as depicted in Figure 5.3 on the left.

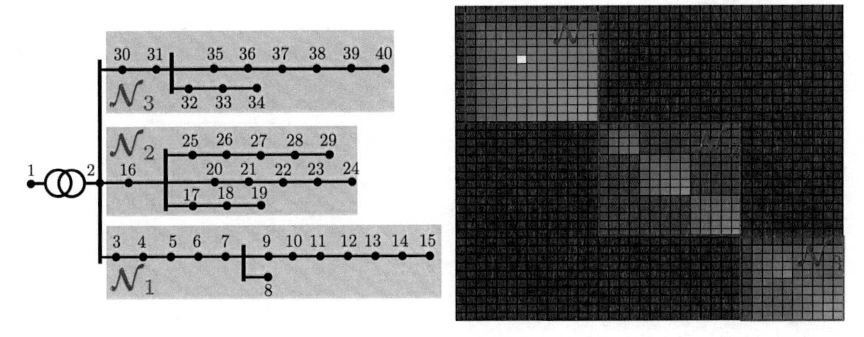

Figure 5.3: 40 node benchmark grid on the left and the corresponding structure of the low voltage grid impedance matrix on the right

This sectioning can be automated based on the coupling strength between the nodes. It amounts in an impedance which describes the reaction strength between two respective leaf nodes as can be seen from the color range in Figure 5.3. The range is from black (matrix entry is zero) over blue up to yellow in case of a high value. Due to the radial structure of the grid, the impedance matrix is block-diagonal. To section the individual sub-grids these blocks can be isolated from the main impedance matrix.

The sub-grids are now smaller than before. This size reduction also has a positive effect on the number of nodes that need to communicate with each other. It should be noted that in turn, it has no influence on the calculation efficiency of the local optimization problem because they do not scale with the size of the sub-grids. However, the approximated power flow only needs to be determined for each section $\mathcal{N}_i$ with the node closest to the root acting as a surrogate slack bus. So the local construction of the power flow approximation now requires the inversion of smaller matrices which increases the calculation speed.

Jacobi type distributed optimization algorithm

There are many methods found in the literature to solve a separable optimization problem in a distributed fashion [NND11]. Again the need for feasible iterations in this thesis limits the amount of possible algorithms to primal decomposition approaches. Jacobi type algorithms are capable to solve a separable optimization problem that is only coupled in the cost function and not in the constraints. The solution of the individual optimization problems can completely run in parallel but need to be synchronized. Following optimization problem is solved in every node for each time instant k

$$\boldsymbol{x}_{k+1}^{[\nu]} = \arg \min_{\boldsymbol{x}^{[\nu]} \in \mathcal{X}^{[\nu]}} f\left(\boldsymbol{x}_k^{[1]}, \cdots, \boldsymbol{x}_k^{[\nu-1]}, \boldsymbol{x}^{[\nu]}, \boldsymbol{x}_k^{[\nu+1]}, \boldsymbol{x}_k^{[N_i]}\right). \tag{5.5}$$

While the variable of the current subsystem ν is calculated from the optimization problem as $\boldsymbol{x}_{k+1}^{[\nu]}$ the other subsystem variables $\{\boldsymbol{x}_k^{[1]}, \cdots, \boldsymbol{x}_k^{[N_i]}\}$ are known – have been communicated by subsystem $\mu = \{1, \cdots, N_i\}$ – and stay constant. After each subsystem solved the local optimization problem, it transmits its solution $\boldsymbol{x}_{k+1}^{[\nu]}$ to the others. There are several update laws to guaranty convergence properties and optimality for this procedure. In this thesis a formulation based on the proposed update laws from [Ven06, RS08] is used. It has the following form

$$\begin{aligned} \tilde{\boldsymbol{x}}_k^{[\nu]} &= \arg \min_{\boldsymbol{x}^{[\nu]} \in \mathcal{X}^{[\nu]}} f\left(\boldsymbol{x}_k^{[1]}, \cdots, \boldsymbol{x}_k^{[\nu-1]}, \boldsymbol{x}^{[\nu]}, \boldsymbol{x}_k^{[\nu+1]}, \boldsymbol{x}_k^{[N]}\right) \\ \boldsymbol{x}_{k+1}^{[\nu]} &= \alpha^{[\nu]} \tilde{\boldsymbol{x}}_k^{[\nu]} + (1 - \alpha^{[\nu]}) \boldsymbol{x}_k^{[\nu]} \end{aligned} \tag{5.6}$$

where $\alpha^{[\nu]}$ are positive weights, that sum up to 1 over all $\nu \in \mathcal{N}$.

5.2 Distributed three-phase balancing algorithm

5.2.1 Local node and constraint setup

Controlled nodes now have some combination of inverter interfaced stationary battery storage and single- as well three-phase chargers for electric vehicles. Single-phase chargers are capable of providing reactive power, as well reducing their active power during charging. Single-phase photo-voltaic units, which are present in the node have a fixed active power in-feed and are treated as a negative load. They are, together with the domestic load, the main source of fixed unbalance in the power grid. The three-phase inverters from the stationary battery and photo-voltaic units are assumed to be controllable with unequal apparent power in the three phases, but have no neutral conductor connection. This consequently means that they can only shift power from one phase to the other but cannot directly influence the neutral conductor current. Consequentially, they can only mitigate effects related to the negative-sequence voltage. Furthermore, the balancing controller cannot reduce or increase the mean active power of the generation units, thus the output which was set by the MPC must stay constant. In summary the local setup used for the balancing control is depicted in Figure 5.4. All relevant variables

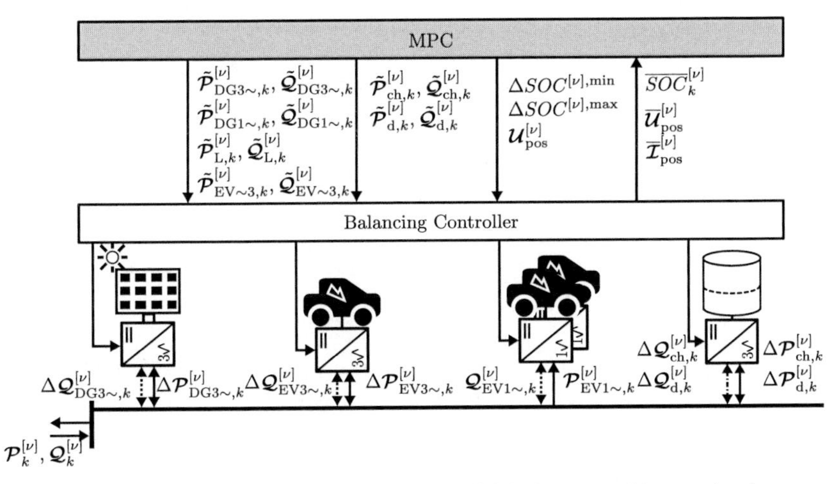

Figure 5.4: Extended local node setup with controllable devices and interaction between MPC and balancing controller

visible in Figure 5.4 will be introduced in the following sections.

Local power balance in the three-phase setup

The active and reactive node power balance from the connected units inside the node with the grid can be formulated as

$$\begin{aligned}\boldsymbol{\mathcal{P}}_k^{[\nu]} =&\tilde{\boldsymbol{\mathcal{P}}}_{\mathrm{L},k}^{[\nu]} + \tilde{\boldsymbol{\mathcal{P}}}_{\mathrm{DG3}\sim,k}^{[\nu]} + \tilde{\boldsymbol{\mathcal{P}}}_{\mathrm{DG1}\sim,k}^{[\nu]} + \tilde{\boldsymbol{\mathcal{P}}}_{\mathrm{EV3}\sim,k}^{[\nu]} + \tilde{\boldsymbol{\mathcal{P}}}_{\mathrm{ch},k}^{[\nu]} + \tilde{\boldsymbol{\mathcal{P}}}_{\mathrm{d},k}^{[\nu]} + \\ &\Delta\boldsymbol{\mathcal{P}}_{\mathrm{DG3}\sim,k}^{[\nu]} + \Delta\boldsymbol{\mathcal{P}}_{\mathrm{EV3}\sim,k}^{[\nu]} + \boldsymbol{\mathcal{P}}_{\mathrm{EV1}\sim,k}^{[\nu]} + \Delta\boldsymbol{\mathcal{P}}_{\mathrm{ch},k}^{[\nu]} + \Delta\boldsymbol{\mathcal{P}}_{\mathrm{d},k}^{[\nu]} \end{aligned} \tag{5.7a}$$

$$\begin{aligned}\boldsymbol{\mathcal{Q}}_k^{[\nu]} =&\tilde{\boldsymbol{\mathcal{Q}}}_{\mathrm{L},k}^{[\nu]} + \tilde{\boldsymbol{\mathcal{Q}}}_{\mathrm{DG3}\sim,k}^{[\nu]} + \tilde{\boldsymbol{\mathcal{Q}}}_{\mathrm{DG1}\sim,k}^{[\nu]} + \tilde{\boldsymbol{\mathcal{Q}}}_{\mathrm{EV3}\sim,k}^{[\nu]} + \tilde{\boldsymbol{\mathcal{Q}}}_{\mathrm{ch},k}^{[\nu]} + \tilde{\boldsymbol{\mathcal{Q}}}_{\mathrm{d},k}^{[\nu]} + \\ &\Delta\boldsymbol{\mathcal{Q}}_{\mathrm{DG3}\sim,k}^{[\nu]} + \Delta\boldsymbol{\mathcal{Q}}_{\mathrm{EV3}\sim,k}^{[\nu]} + \boldsymbol{\mathcal{Q}}_{\mathrm{EV1}\sim,k}^{[\nu]} + \Delta\boldsymbol{\mathcal{Q}}_{\mathrm{ch},k}^{[\nu]} + \Delta\boldsymbol{\mathcal{Q}}_{\mathrm{d},k}^{[\nu]}. \end{aligned} \tag{5.7b}$$

with EV3$\sim$ denoting the power exchange from three-phase and EV1$\sim$ from single-phase electric vehicle chargers inside the node. The indexing for distributed generation is the same. Note, that now they are both variables that are fixed from the MPC, denoted with a tilde symbol and quantities that can be adjusted with the balancing controller shown as Δ values. One exception is the single-phase electric vehicle charger because it's charging power is only influenced by the balancing controller. The fixed quantities (5.7a), (5.7b) are equal to the power balance for the operation strategy with MPC (4.1), (4.2). All fixed variables transmitted from the MPC only have an informative character. They have no influence on the balancing algorithm but need to be considered in the local constraint construction.

Flexible power resources from phase exchange

The variable resources from photo-voltaic units and EV3$\sim$ chargers that are connected via three-phase inverters split into the following phase variables

$$\Delta\boldsymbol{\mathcal{P}}_{\mathrm{DG3}\sim,k}^{[\nu]} = \left(\Delta p_{\mathrm{DG3}\sim,k}^{[\nu],\mathrm{a}} \quad \Delta p_{\mathrm{DG3}\sim,k}^{[\nu],\mathrm{b}} \quad \Delta p_{\mathrm{DG3}\sim,k}^{[\nu],\mathrm{c}}\right)^{\mathrm{T}} \tag{5.8a}$$

$$\Delta\boldsymbol{\mathcal{Q}}_{\mathrm{DG3}\sim,k}^{[\nu]} = \left(\Delta q_{\mathrm{DG3}\sim,k}^{[\nu],\mathrm{a}} \quad \Delta q_{\mathrm{DG3}\sim,k}^{[\nu],\mathrm{b}} \quad \Delta q_{\mathrm{DG3}\sim,k}^{[\nu],\mathrm{c}}\right)^{\mathrm{T}} \tag{5.8b}$$

where $\Delta\boldsymbol{\mathcal{P}}_{\mathrm{DG3}\sim,k}$ and $\Delta\boldsymbol{\mathcal{Q}}_{\mathrm{DG3}\sim,k}$ are carryover variables. It is assumed that the power output of the 3 $\sim$ inverters can be adjusted in the range of $\pm 2\,\%$ of the current active power $\hat{\boldsymbol{\mathcal{P}}}_{\mathrm{DG3}\sim,k}$, $\hat{\boldsymbol{\mathcal{P}}}_{\mathrm{EV3}\sim,k}$ which is exchanged with the grid by the command of the MPC. The reactive power $\Delta\boldsymbol{\mathcal{Q}}_{\mathrm{DG3}\sim,k}$ can be adjusted in the same way as long as the limitation of inverters are not exceeded.
Its value is determined based on the controlled reactive power $\tilde{\boldsymbol{\mathcal{Q}}}_{\mathrm{DG3}\sim,k}$, $\tilde{\boldsymbol{\mathcal{Q}}}_{\mathrm{EV3}\sim,k}$ which is used in the MPC for reactive power support. As mentioned before this flexible power can be interpreted as a redistribution between the phases of the inverter. This means that the PV generation units and three-phase chargers do not inject or draw more power

from the grid than is currently available or needed. To include this balance in the three phases the following equality constraint is introduced

$$0 = \Delta p^{[\nu],\mathrm{a}}_{\mathrm{DG3}\sim,k} + \Delta p^{[\nu],\mathrm{b}}_{\mathrm{DG3}\sim,k} + \Delta p^{[\nu],\mathrm{c}}_{\mathrm{DG3}\sim,k} \tag{5.9a}$$

$$0 = \Delta q^{[\nu],\mathrm{a}}_{\mathrm{DG3}\sim,k} + \Delta q^{[\nu],\mathrm{b}}_{\mathrm{DG3}\sim,k} + \Delta q^{[\nu],\mathrm{c}}_{\mathrm{DG3}\sim,k} \tag{5.9b}$$

which is included for both three-phase charger and photo-voltaic units.

Neutral current injection constraints from inverters

Two-level three-phase inverters are currently still the dominating technology used in low voltage grids. It is possible to use these inverters to generate zero-sequence voltages, but they do not drive a neutral current into the grid because there is no physical return path in the form of a connected conductor [Ye00]. To account for this limitation an additional equality constraint is implemented, forcing the neutral conductor current of the inverter interface $\underline{i}^{\mathrm{n}}_{\mathrm{DG}}$ defined by (2.4) to be zero. The constraints are shown as an example for the variable resource $\Delta\boldsymbol{\mathcal{P}}^{[\nu]}_{\mathrm{DG3}\sim,k}, \Delta\boldsymbol{\mathcal{Q}}^{[\nu]}_{\mathrm{DG3}\sim,k}$ but must be applied to every three-phase inverter.

$$\begin{pmatrix} \overline{\epsilon}^{\mathrm{a,Re}}_{\nu} & \overline{\epsilon}^{\mathrm{b,Re}}_{\nu} & \overline{\epsilon}^{\mathrm{c,Re}}_{\nu} \\ -\overline{\epsilon}^{\mathrm{a,Im}}_{\nu} & -\overline{\epsilon}^{\mathrm{b,Im}}_{\nu} & -\overline{\epsilon}^{\mathrm{c,Im}}_{\nu} \end{pmatrix} \Delta\boldsymbol{\mathcal{P}}^{[\nu]}_{\mathrm{DG3}\sim,k} + \begin{pmatrix} -\overline{\epsilon}^{\mathrm{a,Im}}_{\nu} & -\overline{\epsilon}^{\mathrm{b,Im}}_{\nu} & -\overline{\epsilon}^{\mathrm{c,Im}}_{\nu} \\ -\overline{\epsilon}^{\mathrm{a,Re}}_{\nu} & -\overline{\epsilon}^{\mathrm{b,Re}}_{\nu} & -\overline{\epsilon}^{\mathrm{c,Re}}_{\nu} \end{pmatrix} \Delta\boldsymbol{\mathcal{Q}}^{[\nu]}_{\mathrm{DG3}\sim,k} = \mathbf{0} \tag{5.10}$$

$\overline{\epsilon}^{\xi,\mathrm{Re}}_{\nu}$, $\overline{\epsilon}^{\xi,\mathrm{Im}}_{\nu}\,\forall\xi \in \{\mathrm{a,b,c}\}$ represent the real and imaginary part of the inverse measured nodal voltages in all three phases from the current control cycle. It should be noted, that it is necessary to use this formulation, because the power (5.9a), (5.9b) and current balance (5.10) would be equal if the no-load voltage $\overline{\epsilon}^{\xi,\mathrm{Re}}_{0}$, $\overline{\epsilon}^{\xi,\mathrm{Im}}_{0}$ would be used.

Battery storage constraints

The balancing controller and the MPC share a common battery storage. But, because the battery can undergo extensive usage in the optimal dispatch strategy, only a part of the capacity can be provided to the balancing controller. This part is evaluated by the upper layer MPC and communicated as a fixed limitation. The available change in the state of charge $\Delta SOC^{[\nu]}$ of the battery storage at node ν and the limits are then formulated as a constraint on the charging power as

$$\Delta SOC^{\mathrm{min},[\nu]} \leq \eta^{[\nu]}_{\mathrm{c}} T_{\mathrm{s}} \begin{pmatrix} 1 & 1 & 1 \end{pmatrix} \Delta\boldsymbol{\mathcal{P}}^{[\nu]}_{\mathrm{ch},k} + \frac{T_{\mathrm{s}}}{\eta^{[\nu]}_{\mathrm{d}}} \begin{pmatrix} 1 & 1 & 1 \end{pmatrix} \Delta\boldsymbol{\mathcal{P}}^{[\nu]}_{\mathrm{d},k} \leq \Delta SOC^{\mathrm{max},[\nu]}. \tag{5.11a}$$

Note, that a usage of the battery storage is reported back to the MPC in form of a measurement, as it needs to know the correct state of charge for the operation strategy.

At some instances i.e. when the battery is completely depleted or fully charged, one of the constraints $\Delta SOC^{\max,[\nu]}$, $\Delta SOC^{\min,[\nu]}$ is zero. The sample time and efficiency is equal to the balanced case (4.3).

Inverter capacity constraints

Every component in the nodes $\nu \in \mathcal{N}$, $\xi \in \{\mathrm{a}, \mathrm{b}, \mathrm{c}\}$ that is interfaced to the grid with an inverter has a specific apparent power constraint which is given by the rated power of the inverter. The power output limitation is formulated as a quadratic constraint in the form of a squared euclidean norm. However, in case of the balancing controller, the apparent power constraint is no longer a pure constructional feature of the inverter. Both MPC and the balancing controller share the same resource, to interact with the grid. The relationship between inverter constraints from (4.5b) and the local flexibility from (5.7a), (5.7b) is depicted in Figure 5.5. The constraints can be formulated as

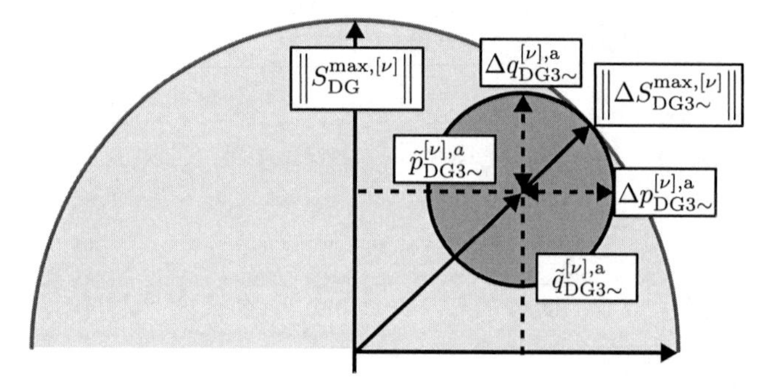

Figure 5.5: Determination of the inverter constraints for the flexible apparent power resources, exemplary for phase **a**

$$\left\|\begin{pmatrix}\Delta p_{\mathrm{ch},k}^{[\nu],\xi}\\ \Delta q_{\mathrm{ch},k}^{[\nu],\xi}\end{pmatrix}\right\|^2 \leq \left(\Delta S_{\mathrm{ch}}^{\max,[\nu]}\right)^2, \left\|\begin{pmatrix}\Delta p_{\mathrm{d},k}^{[\nu],\xi}\\ \Delta q_{\mathrm{d},k}^{[\nu],\xi}\end{pmatrix}\right\|^2 \leq \left(\Delta S_{\mathrm{d}}^{\max,[\nu]}\right)^2, \left\|\begin{pmatrix}\Delta p_{\mathrm{DG3\sim},k}^{[\nu],\xi}\\ \Delta q_{\mathrm{DG3\sim},k}^{[\nu],\xi}\end{pmatrix}\right\|^2 \leq \left(\Delta S_{\mathrm{DG3\sim}}^{\max,[\nu]}\right)^2,$$
$$\left\|\begin{pmatrix}\Delta p_{\mathrm{EV3\sim},k}^{[\nu],\xi}\\ \Delta q_{\mathrm{EV3\sim},k}^{[\nu],\xi}\end{pmatrix}\right\|^2 \leq \left(\Delta S_{\mathrm{EV3\sim}}^{\max,[\nu]}\right)^2 \tag{5.12}$$

$$\Delta \boldsymbol{\mathcal{P}}_{\mathrm{d},k}^{[\nu]} \leq \mathbf{0},\ \mathbf{0} \leq \Delta \boldsymbol{\mathcal{P}}_{\mathrm{ch},k}^{[\nu]}. \tag{5.13}$$

Electric vehicles that are connected to single-phase chargers are assumed to be controllable in their whole power range. This means that the active power $p_{\mathrm{EV1\sim},k}^{[\nu],\xi}$ and the corresponding reactive power $q_{\mathrm{EV1}\sim,k}^{[\nu],\xi}$ of every connected charger is adjustable from the current charging power down to zero. Because it is known to which phase $\xi \in \{\mathrm{a,b,c}\}$ the chargers is connected in each node ν, the constraints can be formulated individually as

$$\left\|\begin{pmatrix} p_{\mathrm{EV1}\sim,k}^{[\nu],\mathrm{a}} \\ q_{\mathrm{EV1}\sim,k}^{[\nu],\mathrm{a}} \end{pmatrix}\right\|^2 \leq \left(S_{\mathrm{EV1}\sim,k}^{\mathrm{max,a},[\nu]}\right)^2, \left\|\begin{pmatrix} p_{\mathrm{EV1}\sim,k}^{[\nu],\mathrm{b}} \\ q_{\mathrm{EV1}\sim,k}^{[\nu],\mathrm{b}} \end{pmatrix}\right\|^2 \leq \left(S_{\mathrm{EV1}\sim,k}^{\mathrm{max,b},[\nu]}\right)^2, \left\|\begin{pmatrix} p_{\mathrm{EV1}\sim,k}^{[\nu],\mathrm{c}} \\ q_{\mathrm{EV1}\sim,k}^{[\nu],\mathrm{c}} \end{pmatrix}\right\|^2 \leq \left(S_{\mathrm{EV1}\sim,k}^{\mathrm{max,c},[\nu]}\right)^2 \quad (5.14)$$

$$\boldsymbol{\mathcal{P}}_{\mathrm{EV1}\sim,k}^{[\nu]} \leq \mathbf{0} \quad (5.15)$$

The quadratic constraints (5.12), (5.14) are summarized as $\mathbb{S}_{\mathrm{loc},k}^{[\nu]}$

5.2.2 Coordination of the distributed balancing controllers

The following section introduces the coordination strategy and distributed solution of the unbalance mitigation problem. The individual steps of the algorithm are detailed in the different sections.

Operating point determination for the approximated power flow

In a first step every node in the grid that is equipped with measurement devices determines the local voltage magnitude $\left\|\overline{\underline{u}}_{\nu}^{\xi}\right\|$ and apparent power $\overline{\boldsymbol{\mathcal{S}}}_k^{[\nu]}$ injection in every phase $\xi = \{\mathrm{a,b,c}\}$ for time step k. Note that

$$\overline{\boldsymbol{\mathcal{S}}}_k^{[\nu]} = \begin{pmatrix} \overline{p}_k^{[\nu],\mathrm{a}} & \overline{p}_k^{[\nu],\mathrm{b}} & \overline{p}_k^{[\nu],\mathrm{c}} & \overline{q}_k^{[\nu],\mathrm{a}} & \overline{q}_k^{[\nu],\mathrm{b}} & \overline{q}_k^{[\nu],\mathrm{c}} \end{pmatrix}^{\mathrm{T}}$$

is a combination of fixed controlled apparent power from the balanced controller (MPC) and an unknown but measurable apparent power $\Delta\overline{\boldsymbol{\mathcal{S}}}_k^{[\nu]}$. The difference is mainly caused by single-phase components both on generation and on load side.

The voltage magnitudes are then mapped onto the three phases based on (5.24) to be available in magnitude and phase. After processing the collected measurements, all nodes in the grid broadcast their own operating point $\overline{\boldsymbol{\mathcal{S}}}_k^{[\nu]}, \overline{\boldsymbol{\mathcal{U}}}_k^{[\nu]}$ onto the common communication bus. Moreover, each controlled node ν receive $\overline{\boldsymbol{\mathcal{S}}}_k^{[\kappa]}, \overline{\boldsymbol{\mathcal{U}}}_k^{[\kappa]}$ from the other controllers and $\overline{\boldsymbol{\mathcal{S}}}_k^{[\mu]}, \overline{\boldsymbol{\mathcal{U}}}_k^{[\mu]}$ from uncontrolled nodes in the grid.
The collected measurements are then used by the local controllers to parameterize (2.29) and to determine the power flow approximation $\mathbb{A}_{\mathrm{PF}}^{-1}\mathbb{B}_{\mathrm{PF}}$ which is valid for the current operating point. Note that this is done for each of the sections from Figure 5.3 using

the node closest to the substation transformer as the surrogate slack bus. This node delivers the voltage $\overline{\mathcal{U}}_{0,k}^{[\mathcal{N}_i]}$. Usually, the node closest to the substation transformer only has a small variation in voltage magnitude and phase angle from the substation itself. Thus, these assumptions introduce only a small error.
Furthermore, the approximation (2.29) and the decomposition matrices (2.34) are determined separately for each section $\mathcal{N}_i$. The power flow approximation for each section takes on the following form

$$\mathcal{U}_k^{[\mathcal{N}_i]} = \mathbb{C}_{\mathrm{PF}}^{[\mathcal{N}_i]} \left(\mathbb{A}_{\mathrm{PF}}^{[\mathcal{N}_i]}\right)^{-1} \mathbb{B}_{\mathrm{PF}}^{[\mathcal{N}_i]} \overline{\mathcal{S}}_k^{[\mathcal{N}_i]} + \overline{\mathcal{U}}_{0,k}^{[\mathcal{N}_i]}, \; i \in \{1,2,3\}.$$

The amount of nodes inside of the grid sections $\mathcal{N}_i$, as depicted in Figure 5.3 is denoted as N_i.

Sequence-voltage influence of controlled and uncontrolled nodes

Only a few nodes $\nu, \kappa \in \mathcal{M}_i \subset \mathcal{N}_i$ in the grid are equipped with a controller and uncontrolled nodes $\mu \notin \mathcal{M}_i$ as well as the unbalance they introduce into the grid must be considered separately. However, as the power flow approximation is linear, the controlled nodes are able to recover the negative-sequence (5.16) and zero-sequence (5.17) components from the measurement broadcast for the individual uncontrolled nodes. Each component can be determined as a summation of the following form

$$\Delta\mathcal{U}_{\mathrm{neg},k}^{[\mathcal{N}_i,\mu\notin\mathcal{M}_i]} = \sum_{\mu\in\mathcal{N}_i,\mu\notin\mathcal{M}_i} \underbrace{\sum_{\xi\in\{\mathrm{a,b,c}\}} \left(\mathbb{M}_{\mathrm{neg}}^{[\mathcal{N}_i]} \mathbb{K}_{\mathrm{PF}}^{[\mathcal{N}_i]} \mathbb{C}_{\mathrm{PF}}^{[\mathcal{N}_i]} \left(\mathbb{A}_{\mathrm{PF}}^{[\mathcal{N}_i]}\right)^{-1} \mathbb{B}_{\mathrm{PF}}^{[\mathcal{N}_i]} \underbrace{\begin{pmatrix} \mathbb{1}_\mu^\xi & \mathbf{0} \\ \mathbf{0} & \mathbb{1}_\mu^\xi \end{pmatrix}}_{\mathbb{2}_\mu^\xi} \right) \begin{pmatrix} \overline{p}_k^{[\mu],\xi} \\ \overline{q}_k^{[\mu],\xi} \end{pmatrix}}_{\mathbb{H}_{\mathrm{PF,neg}}^{[\mu]} \overline{\mathcal{S}}_k^{[\mu]}} \tag{5.16}$$

$$\Delta\mathcal{U}_{\mathrm{zero},k}^{[\mathcal{N}_i,\mu\notin\mathcal{M}_i]} = \sum_{\mu\in\mathcal{N}_i,\mu\notin\mathcal{M}_i} \underbrace{\sum_{\xi\in\{\mathrm{a,b,c}\}} \left(\mathbb{M}_{\mathrm{zero}}^{[\mathcal{N}_i]} \mathbb{K}_{\mathrm{PF}}^{[\mathcal{N}_i]} \mathbb{C}_{\mathrm{PF}}^{[\mathcal{N}_i]} \left(\mathbb{A}_{\mathrm{PF}}^{[\mathcal{N}_i]}\right)^{-1} \mathbb{B}_{\mathrm{PF}}^{[\mathcal{N}_i]} \begin{pmatrix} \mathbb{1}_\mu^\xi & \mathbf{0} \\ \mathbf{0} & \mathbb{1}_\mu^\xi \end{pmatrix} \right) \begin{pmatrix} \overline{p}_k^{[\mu],\xi} \\ \overline{q}_k^{[\mu],\xi} \end{pmatrix}}_{\mathbb{H}_{\mathrm{PF,zero}}^{[\mu]} \overline{\mathcal{S}}_k^{[\mu]}}. \tag{5.17}$$

The formulation is shown as an example for the uncontrolled part but can be applied to the controlled nodes in the same manner. The summation for each sequence-component is formulated as an equivalent matrix $\mathbb{H}_{\mathrm{PF,neg}}^{[\mu]}$ and $\mathbb{H}_{\mathrm{PF,zero}}^{[\mu]}$ with size $2N_i \times 6$ that maps the measured nodal power injection $\overline{\mathcal{S}}_k^{[\mu]}$ of the uncontrolled nodes μ onto the negative and zero-sequence component of every node inside the section respectively.

On-line constraint determination of single-phase electric vehicle chargers

The local constraints for the balancing controllers must be determined before the mitigation procedure starts operating. The prediction of single-phase PV unit in-feed is assumed to be the same as from the balanced case. Even though the position of PV plants has an effect due to the difference in radiation during cloud coverage changes [UT13], it is further assumed that this difference does not play a significant role as mentioned in [Ker10]. In order to determine the power constraints on the single-phase electric vehicle chargers some form of exchange protocol is assumed to be present that tells the balancing controller the power rating of the charger and the phase, the vehicle is connected to. The battery of the car is not explicitly considered for optimization. It is assumed that the power rating does not overcharge the battery because the supervisory system of the electric vehicle communicates the maximum charging profile of the battery to the balancing controller.

Cost function and constraints of the optimization problem

As has been stated before the unbalance defined by λ_ν^{neg} and $\lambda_\nu^{\text{zero}}$ can be reduced by only minimizing the relative negative- and zero-sequence component of the individual nodes that have been depicted in Figure 5.2 for node $\nu = 1$ and defined in (5.4c)-(5.4d). The controlled nodes do this by solving an optimization problem that again consists of two parts. A part that describes the interaction of the local controller with the grid $\mathcal{J}_\text{G}$ and a part that optimizes the node internal powers $\mathcal{J}_\text{I}$. The optimization problem describing the grid interaction includes the cost for the unbalances as a squared euclidean norm of the undesired sequence-voltage in the grid.

$$
\begin{aligned}
\mathcal{J}_\text{G}^{[\nu]} = & \underbrace{\left\| \Delta \mathcal{U}_{\text{neg},k}^{[\mathcal{N}_i,\nu\in\mathcal{M}_i]} \right\|^2 + \left\| \Delta \mathcal{U}_{\text{zero},k}^{[\mathcal{N}_i,\nu\in\mathcal{M}_i]} \right\|^2}_{\text{Controlled unbalance } \nu} + \underbrace{\left\| \Delta \mathcal{U}_{\text{neg},k}^{[\mathcal{N}_i,\kappa\in\mathcal{M}_i]} \right\|^2 + \left\| \Delta \mathcal{U}_{\text{zero},k}^{[\mathcal{N}_i,\kappa\in\mathcal{M}_i]} \right\|^2}_{\text{Controlled unbalance } \kappa} + \\
& \underbrace{\left\| \Delta \mathcal{U}_{\text{neg},k}^{[\mathcal{N}_i,\mu\notin\mathcal{M}_i]} \right\|^2 + \left\| \Delta \mathcal{U}_{\text{zero},k}^{[\mathcal{N}_i,\mu\notin\mathcal{M}_i]} \right\|^2}_{\text{Unbalance from uncontrolled nodes } \mu}
\end{aligned}
\tag{5.18}
$$

This more general cost function is now distributed to every local controller keeping in mind that every grid section $\mathcal{N}_i$ has several controlled and uncontrolled nodes. Using the approach (5.16) and (5.17) to reformulate the controlled nodes influence of the sequence-voltage and substitute into (5.18) leads to

$$\begin{aligned}\mathcal{J}_{\mathrm{G}}^{[\nu]} = &\left(\boldsymbol{\mathcal{S}}_k^{[\nu]}\right)^{\mathrm{T}} \begin{pmatrix}\mathbb{H}_{\mathrm{PF,neg}}^{[\nu]}\\ \mathbb{H}_{\mathrm{PF,zero}}^{[\nu]}\end{pmatrix}^{\mathrm{T}} \begin{pmatrix}\mathbb{H}_{\mathrm{PF,neg}}^{[\nu]}\\ \mathbb{H}_{\mathrm{PF,zero}}^{[\nu]}\end{pmatrix} \boldsymbol{\mathcal{S}}_k^{[\nu]} + \\ &2\begin{pmatrix}\sum_{\kappa\in\mathcal{M}_i} \mathbb{H}_{\mathrm{PF,neg}}^{[\kappa]}\boldsymbol{\mathcal{S}}_{j-1,k}^{[\kappa]} + \sum_{\mu\notin\mathcal{M}_i} \mathbb{H}_{\mathrm{PF,neg}}^{[\mu]}\overline{\boldsymbol{\mathcal{S}}}_k^{[\mu]}\\ \sum_{\kappa\in\mathcal{M}_i} \mathbb{H}_{\mathrm{PF,zero}}^{[\kappa]}\boldsymbol{\mathcal{S}}_{j-1,k}^{[\kappa]} + \sum_{\mu\notin\mathcal{M}_i} \mathbb{H}_{\mathrm{PF,zero}}^{[\mu]}\overline{\boldsymbol{\mathcal{S}}}_k^{[\mu]}\end{pmatrix}^{T} \begin{pmatrix}\mathbb{H}_{\mathrm{PF,neg}}^{[\nu]}\\ \mathbb{H}_{\mathrm{PF,zero}}^{[\nu]}\end{pmatrix} \boldsymbol{\mathcal{S}}_k^{[\nu]}\end{aligned} \tag{5.19}$$

with

- The influence of other controlled nodes $\sum_{\kappa\in\mathcal{M}_i} \mathbb{H}_{\mathrm{PF,neg}}^{[\kappa]}\boldsymbol{\mathcal{S}}_{j-1,k}^{[\kappa]}$ with j marking the iteration and thus the version of the controlled node power of controller κ.
- The influence of other uncontrolled nodes $\sum_{\mu\notin\mathcal{M}_i} \mathbb{H}_{\mathrm{PF,neg}}^{[\nu]}\overline{\boldsymbol{\mathcal{S}}}_k^{[\mu]}$.
- The vector of apparent power from the grid interaction of node ν, which is the sum of all internal power $\boldsymbol{\mathcal{S}}_k^{[\nu]} = \left[p_k^{[\nu],\mathrm{a}}\ p_k^{[\nu],\mathrm{b}}\ p_k^{[\nu],\mathrm{c}}\ q_k^{[\nu],\mathrm{a}}\ q_k^{[\nu],\mathrm{b}}\ q_k^{[\nu],\mathrm{c}}\right]^{\mathrm{T}}$.

The linear part of the cost function $\mathcal{J}_I^{[\nu]}$ penalizes all the internal flexible powers from battery storage, electric vehicles and distributed generation and ensures the constraints which have been introduced in the beginning of this section. The local cost function takes on the following form

$$\mathcal{J}_I^{[\nu]} = \left(\begin{pmatrix}1 & 1 & 1\end{pmatrix} \otimes \left(C_{\mathrm{DG3}\sim}^{[\nu]}\ \ C_{\mathrm{EV3}\sim}^{[\nu]}\ \ C_{\mathrm{EV1}\sim}^{[\nu]}\ \ C_{\mathrm{ch}}^{[\nu]}\ \ C_{\mathrm{d}}^{[\nu]}\right)\begin{pmatrix}\mathbf{0} & \mathbf{0} & \mathbf{0} & \mathbf{0}\end{pmatrix}\right) \boldsymbol{\mathcal{S}}_{\mathrm{loc},k}^{[\nu]} \tag{5.20}$$

with

$$\boldsymbol{\mathcal{S}}_{\mathrm{loc},k}^{[\nu]} = \Bigg[\underbrace{\Delta\boldsymbol{\mathcal{P}}_{\mathrm{DG3}\sim,k}^{[\nu],\mathrm{T}}\, \Delta\boldsymbol{\mathcal{P}}_{\mathrm{EV3}\sim,k}^{[\nu],\mathrm{T}}\, \boldsymbol{\mathcal{P}}_{\mathrm{EV1}\sim,k}^{[\nu],\mathrm{T}}\, \Delta\boldsymbol{\mathcal{P}}_{\mathrm{ch},k}^{[\nu],\mathrm{T}}\, \Delta\boldsymbol{\mathcal{P}}_{\mathrm{d},k}^{[\nu],\mathrm{T}}}_{\boldsymbol{\mathcal{P}}_{\mathrm{loc},k}^{[\nu]}}\quad \underbrace{\Delta\boldsymbol{\mathcal{Q}}_{\mathrm{DG3}\sim,k}^{[\nu],\mathrm{T}}\, \Delta\boldsymbol{\mathcal{Q}}_{\mathrm{EV3}\sim,k}^{[\nu],\mathrm{T}}\, \boldsymbol{\mathcal{Q}}_{\mathrm{EV1}\sim,k}^{[\nu],\mathrm{T}}\, \Delta\boldsymbol{\mathcal{Q}}_{\mathrm{ch},k}^{[\nu],\mathrm{T}}\, \Delta\boldsymbol{\mathcal{Q}}_{\mathrm{d},k}^{[\nu],\mathrm{T}}}_{\boldsymbol{\mathcal{Q}}_{\mathrm{loc},k}^{[\nu]}}\Bigg]^{\mathrm{T}}.$$

Distributed solution of the optimization problem

With the additional constraints defined before (5.19)+(5.20) can be classified as a coupled cost decoupled constrained (CCDC) problem and thus be solved via a distributed Jacobi algorithm [NND11]. The core idea of solving the unbalance-mitigation problem in a distributed way with the Jacobi method is to decompose the overall problem into smaller sub-problems which can be solved in parallel by each controller. These solutions happen between two consecutive time steps k and will further be called iterations with the index j as depicted in Figure 5.6.

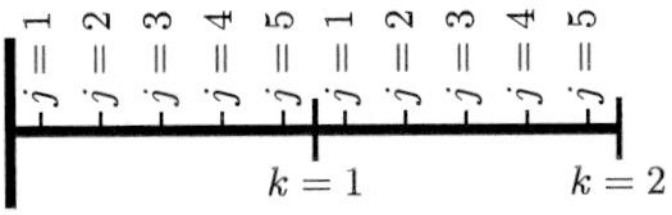

Figure 5.6: Exemplary time axis and iteration index j between two consecutive time steps k in the distributed Jacobi algorithm

The solution vector of the sub-problems are exchanged by the local controllers present in the sub-grid $\mathcal{N}_i$ after every iteration j. The optimization problem that has to be solved by the local controllers takes on the following form

$$\begin{aligned}
&\begin{pmatrix} \check{\boldsymbol{\mathcal{S}}}^{[\nu]}_{j,k} \\ \check{\boldsymbol{\mathcal{S}}}^{[\nu]}_{\text{loc},j,k} \end{pmatrix} = \arg \min_{\boldsymbol{\mathcal{S}}^{[\nu]}_{k}, \boldsymbol{\mathcal{S}}^{[\nu]}_{\text{loc},k}} \quad \mathcal{J}^{[\nu]}_{\text{glo}} + \mathcal{J}^{[\nu]}_{\text{loc}} \\
&\text{s.t.:} \\
&\boldsymbol{\mathcal{S}}^{[\nu]}_{\text{loc},k} \in \mathbb{S}^{[\nu]}_{\text{loc},k} \\
&0 = \begin{pmatrix} 1 & 1 & 1 \end{pmatrix} \Delta \boldsymbol{\mathcal{P}}^{[\nu]}_{\vartheta,k} \\
&0 = \begin{pmatrix} 1 & 1 & 1 \end{pmatrix} \Delta \boldsymbol{\mathcal{Q}}^{[\nu]}_{\vartheta,k} \\
&\boldsymbol{\mathcal{S}}^{[\nu]}_{k} - \begin{pmatrix} \sum \boldsymbol{\mathcal{P}}^{[\nu]}_{\text{loc},k} \\ \sum \boldsymbol{\mathcal{Q}}^{[\nu]}_{\text{loc},k} \end{pmatrix} = -\overline{\boldsymbol{\mathcal{S}}}^{[\nu]}_{k} \\
&\begin{pmatrix} \overline{\epsilon}^{\text{a,Re}}_{\nu} & \overline{\epsilon}^{\text{b,Re}}_{\nu} & \overline{\epsilon}^{\text{c,Re}}_{\nu} \\ -\overline{\epsilon}^{\text{a,Im}}_{\nu} & -\overline{\epsilon}^{\text{b,Im}}_{\nu} & -\overline{\epsilon}^{\text{c,Im}}_{\nu} \end{pmatrix} \Delta \boldsymbol{\mathcal{P}}^{[\nu]}_{\vartheta,k} + \begin{pmatrix} -\overline{\epsilon}^{\text{a,Im}}_{\nu} & -\overline{\epsilon}^{\text{b,Im}}_{\nu} & -\overline{\epsilon}^{\text{c,Im}}_{\nu} \\ -\overline{\epsilon}^{\text{a,Re}}_{\nu} & -\overline{\epsilon}^{\text{b,Re}}_{\nu} & -\overline{\epsilon}^{\text{c,Re}}_{\nu} \end{pmatrix} \Delta \boldsymbol{\mathcal{Q}}^{[\nu]}_{\vartheta,k} = \mathbf{0} \\
&\Delta SOC^{[\nu],\min} \leq \begin{pmatrix} \mathbf{0} & \mathbf{0} & \begin{pmatrix} 1 & 1 & 1 \end{pmatrix} T_{\text{s}} \eta^{[\nu]}_{\text{ch}} & \begin{pmatrix} 1 & 1 & 1 \end{pmatrix} T_{\text{s}} \frac{1}{\eta^{[\nu]}_{\text{d}}} & \mathbf{0} & \mathbf{0} & \mathbf{0} & \mathbf{0} \end{pmatrix} \boldsymbol{\mathcal{S}}^{[\nu]}_{\text{loc},k} \leq \Delta SOC^{[\nu],\max} \\
&\boldsymbol{\mathcal{P}}^{[\nu]}_{\text{EV1}\sim,k} \leq \mathbf{0} \\
&\Delta \boldsymbol{\mathcal{P}}^{[\nu]}_{\text{d},k} \leq \mathbf{0}, \ \mathbf{0} \leq \Delta \boldsymbol{\mathcal{P}}^{[\nu]}_{\text{ch},k}
\end{aligned} \tag{5.21}$$

with $\vartheta = \{\text{EV3} \sim, \text{DG3} \sim, \text{cd}, \text{d}\}$. After this controller ν uses the following update law to receive its own decision variable $\boldsymbol{\mathcal{S}}^{[\nu]}_{j,k}$

$$\boldsymbol{\mathcal{S}}^{[\nu]}_{j,k} = \alpha^{[\nu]} \check{\boldsymbol{\mathcal{S}}}^{[\nu]}_{j,k} + \left(1 - \alpha^{[\nu]}\right) \boldsymbol{\mathcal{S}}^{[\nu]}_{j-1,k}, \ 0 < \alpha^{[\nu]}. \tag{5.22}$$

$\boldsymbol{\mathcal{S}}^{[\nu]}_{j,k}$ is then shared with the others. This process is repeated until following condition holds

$$\left\| \boldsymbol{\mathcal{S}}_{j,k}^{[\nu]} - \boldsymbol{\mathcal{S}}_{j-1,k}^{[\nu]} \right\| \leq \psi_{\breve{\mathcal{S}}}^{[\nu]} \tag{5.23}$$

with the scaling factor $\alpha^{[\nu]}$ with $\nu \in \mathcal{M}_i$ and $\sum_{\nu \in \mathcal{M}_i} \alpha^{[\nu]} = 1$. The scaling factor should be chosen in accord to the amount of balancing power each controller can contribute to the mitigation. Such that a controller with more flexible power receives a higher value of $\alpha^{[\nu]}$. A negotiation of this factor is possible, but in this work it is assumed that it was calculated with a proper method.

The global cost function (5.19) describes the full grid unbalance with relation to positive- and negative-sequence component. $\boldsymbol{\mathcal{S}}_{j,k}^{[\nu]}$ and $\boldsymbol{\mathcal{S}}_{j-1,k}^{[\nu]}$ are the solutions of (5.19) for node ν at iterate j and $j-1$, respectively. Note that (5.22) is a pure local update that needs no negotiation with the others. Furthermore, the apparent power $\boldsymbol{\mathcal{S}}_{j,k}^{[\nu]}$ can be applied to the grid by controller $[\nu]$ for every iteration j of the optimization problem (5.21), because it always satisfies the power flow and the local apparent power constraints.
After the solution of (5.21) the controllers exchange their calculated $\boldsymbol{\mathcal{S}}_{j,k}^{[\nu]}$ with the other controllers for the implementation in the next iteration. The algorithm is terminated if the apparent power change between two consecutive iterations is smaller than some pre-defined threshold $\psi_{\breve{\mathcal{S}}}^{[\nu]}$. This threshold is tested by each local controller before the overall distributed algorithm terminates and after every solution of (5.21). It should be noted that the optimization problem (5.19) is a quadratically constrained quadratic problem (QPQC). QPQCs can be efficiently solved with several MATLAB toolboxes integrating an Interior Point algorithm [CGW05].

To summarize the algorithm proposed in this section, the distributed mitigation is depicted as a flowchart in Figure 5.7.

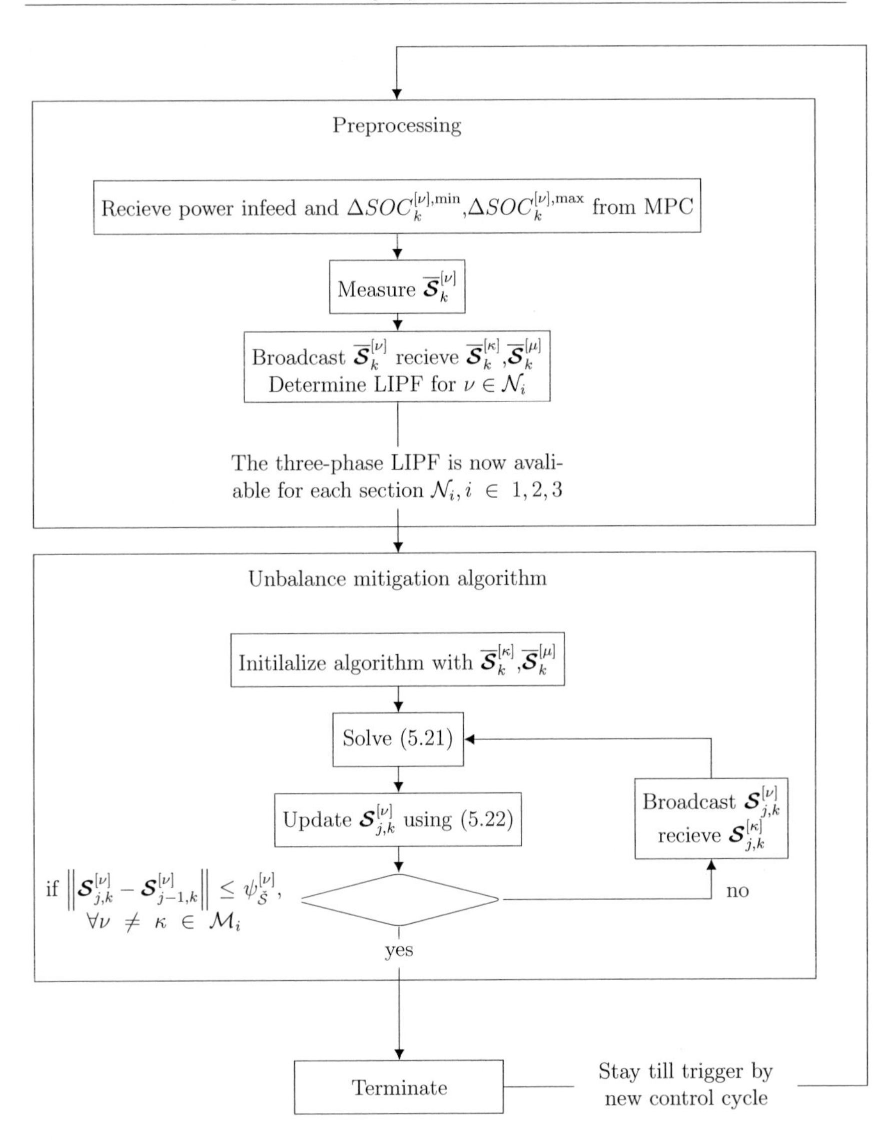

Figure 5.7: Algorithm flowchart depicting every step of the distributed mitigation algorithm in compact form

5.3 Validation of the distributed balancing approach

The distributed control algorithm is validated with the 40 node benchmark grid presented in Chapter 5.2 and the performance is evaluated with regard to the locally available components. As has been pointed out, depending on the generation-unit, battery or EV charger type – different technical possibilities to provide flexible power resources arise. In the following, these are investigated and compared.

It goes without saying that in case of a full controller coverage – each node in the grid is supervised by a controller – the results of the mitigation will be as good as possible. In a realistic setup, this scenario might be infeasible or even unnecessary because not every node in the grid does have the capabilities to participate. Thus, only limited controller coverage is assumed and investigated further. Each section $\mathcal{N}_i$ in the grid as introduced in Figure 5.3 is equipped with three local controllers. Three-phase demand and production data is received from the balanced strategy and based on the results with the setup of Table 4.3. Single-phase data for electric vehicles and small photo-voltaic units can be found in Table B.7 and Table B.8 respectively. The cable values, information about the substation transformer and unbalanced domestic load can be found in the Appendix B, in Table B.5 and Table B.6.

The validation is divided into three parts:

- First the performance of estimating the sequence-voltage based on the LIPF is assessed by comparing it with the values directly calculated from the measurements of a virtual smart-meter and the values directly calculated by OpenDSS.
- The second part investigates the effect of neglecting the ground in the model based on the LIPF on the sequence-voltage estimation. It should be noted that with the LIPF both voltages – phase to neutral and phase to the ground – can be calculated.
- In the third part, the effectiveness of the distributed balancing approach is shown for different possible approaches. The different specific results originating from technical assumptions are discussed.

5.3.1 Uncontrolled sequence-voltage analysis

First the unbalanced low voltage grid introduced in Chapter 3, Figure 3.3 is investigated without deploying any controller. The results of the investigations are depicted in box-plots for better representation. The Box-plot representation in this work has the following properties:

- Colored boxes represent the range in which 95% of determined values lie in.

- The yellow circles inside the colored boxes show the mean value for the 24 h power flow simulation.
- Lines outside the boxes represent the minimum and maximum values that occurred.
- Outliers (values that occur only once during the simulation) are not displayed.

Unbalance assessment: local measurements vs. linear interpolation power flow

Local measurement devices like a smart-meters determine the phase voltage magnitude between line and neutral conductor. The sequence-components of the voltage are then calculated based on these magnitude measurements with (2.28). As the smart-meters do not deliver phase angle information the values have to be transformed by mapping them onto real and imaginary part of the three-phase system with

$$\begin{pmatrix} \underline{\overline{u}}_{\nu}^{\mathrm{a}} \\ \underline{\overline{u}}_{\nu}^{\mathrm{b}} \\ \underline{\overline{u}}_{\nu}^{\mathrm{c}} \end{pmatrix} = \begin{pmatrix} 1 & 0 & 0 \\ 0 & \underline{m} & 0 \\ 0 & 0 & \underline{m}^2 \end{pmatrix} \begin{pmatrix} \|\underline{\overline{u}}_{\nu}^{\mathrm{a}}\| \\ \|\underline{\overline{u}}_{\nu}^{\mathrm{b}}\| \\ \|\underline{\overline{u}}_{\nu}^{\mathrm{c}}\| \end{pmatrix} . \tag{5.24}$$

After this the unbalance factors (5.3a) and (5.3b) can be calculated from the result of (5.24). It is obvious that through the missing phase $\varphi_{\nu}^{\mathrm{a}}, \varphi_{\nu}^{\mathrm{b}}, \varphi_{\nu}^{\mathrm{c}}$ some information is lost with this mapping. The consequence is that the negative- and zero-sequence component are equal in magnitude and conjugate in phase as is depicted for an extreme scenario in Figure 5.8.

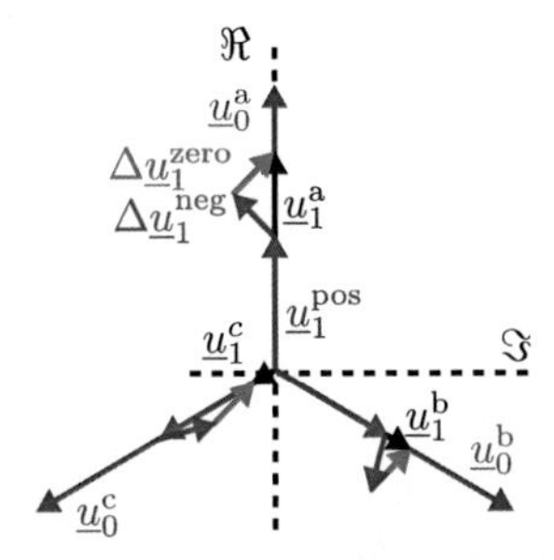

Figure 5.8: Mapping of the magnitude measurement on the respective phases. Left the decomposition in the sequence components and on the right the transformation relationship

A distinction between the two quantities based on the unbalance factors is not possible anymore. This originates from the fact that based on the mapping (5.24) the negative-sequence transformation switches the phases **a** and **b** which does not make a difference regarding the complex-valued summation. In the case of the zero-sequence, the mean value of the three voltages is calculated. In the case of the negative-sequence, the imaginary part is switched in sign. The values are consequently conjugate to each other. The difference between the model-based power flow approximation developed in this work and the direct measurement is demonstrated with the benchmark grid. The OpenDSS result is used as a reference and the unbalance factors (5.3a) and (5.3b) are determined for all three variants as depicted in Figure 5.9

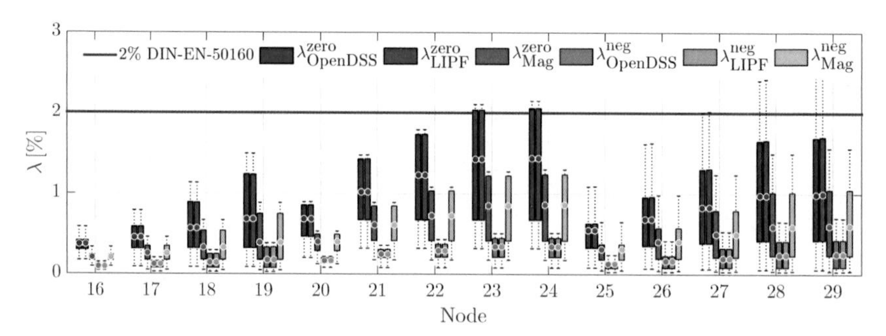

Figure 5.9: Negative- (5.2a) and zero-sequence (5.2b) components from section $\mathcal{N}_2$ in Figure 5.3, from OpenDSS reference, LIPF and from virtual smart-meter magnitude measurement data.

It can be observed from Figure 5.9 that the zero-sequence is under- and the negative-sequence is overestimated when only the magnitude of the voltage is measured. In a feedback-control this would lead to a wrong control error and consequently would not achieve the desired set-point. In summary this means that magnitude information is not sufficient if it is desired to explicitly mitigate the sequence-components of the voltage.

Unbalance assessment: influence of grounding

It has been pointed out in Section 2.2.2 that the effect of regular grounding at different nodes in the grid is neglected, because this represents the worst case scenario regarding the neutral wire voltage. In real applications the grounding impedance can be found to be between $10\,\Omega - 50\,\Omega$ [Ben15] which is quite high compared to the neutral conductor impedance between two consecutive nodes which is only in the range of several hundred mΩ up to $1\,\Omega$.

The blue boxes in Figure 5.10- 5.12 show the result with a neutral conductor and no additional grounding at the individual nodes in λ^{neg}, λ^{zero} and with perfect grounding or a neutral conductor impedance of $0\,\Omega$, as $\lambda_{\text{g}}^{\text{neg}}$, $\lambda_{\text{g}}^{\text{zero}}$. The result in Figure 5.10 - 5.12 show that λ^{zero} changes significantly for the case of no grounding at the individual nodes while the difference in λ^{neg} is quite small.
In case of a neglected ground connection, the neutral conductor has to carry all the return current. The unbalance in the nodes cause a neutral voltage drop, which in turn result in a zero-sequence component in the node voltage. On the other hand, in the case of perfect grounding λ^{zero} is much smaller than in the ungrounded case. Because neutral conductor and ground are physically connected in parallel, the ground carries all the return current which does not change its potential. The missing voltage drop causes a strong reduction in the zero-sequence component of the voltage. In summary, the investigation shows that:

- The zero-sequence component has to be taken into account in the unbalance investigation due to the single-phase generation units and domestic demand in low voltage grids.
- The neutral conductor of the low voltage grid has to be modeled explicitly and considered in the calculation.
- Current flow in the neutral conductor and the resulting offset voltage has nearly no effect on the negative-sequence component calculated at the nodes. This result has also been found by [Ben15].

Moreover, the results depicted in the Figure 5.10 - 5.12 show that under most of the investigated circumstances the zero-sequence and thus the unbalance factor related to it (5.3b) is dominant. The reason for this is the fact that distributed generation units in Europe are either balanced three-phase or single-phase, whereas in the latter case the neutral conductor is used for the return current. On the domestic demand side, the situation is more or less the same [UT13]. In the presented scenario λ^{zero} is mainly influence by EV1$\sim$ chargers and the single-phase PV generation units because the domestic demand is much smaller.

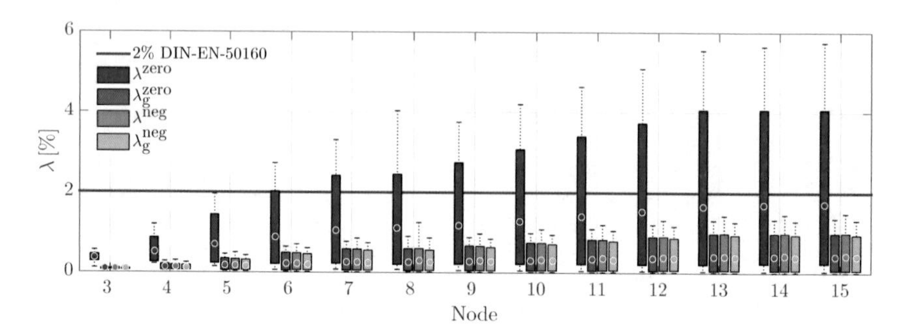

Figure 5.10: Negative- (5.2a) and zero-sequence (5.2b) components from section $\mathcal{N}_1$ in Figure 5.3, without and with perfect grounding

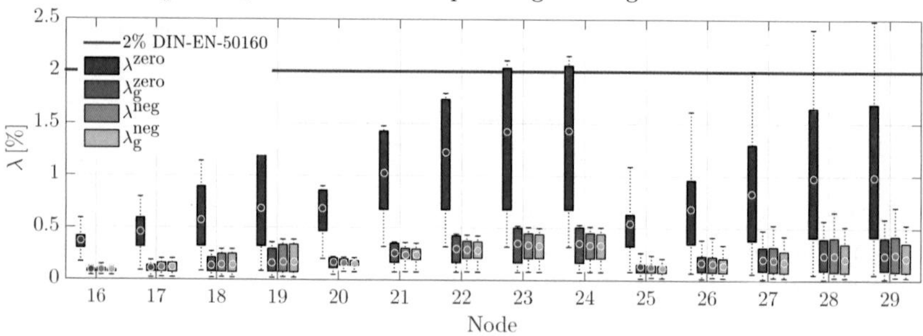

Figure 5.11: Negative- (5.2a) and zero-sequence (5.2b) components from section $\mathcal{N}_2$ in Figure 5.3, without and with perfect grounding

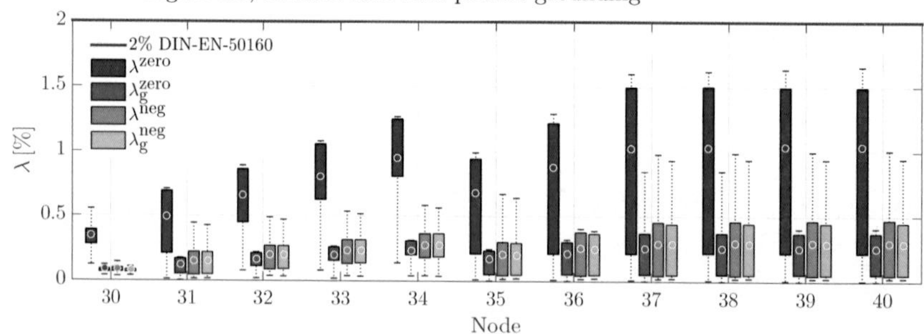

Figure 5.12: Negative- (5.2a) and zero-sequence (5.2b) components from section $\mathcal{N}_3$ in Figure 5.3, without and with perfect grounding

5.3.2 Controlled selective and unified sequence-voltage mitigation

To have a significant influence on the negative-sequence component in the grid the connected load or generation would have to have an unequal phase loading in a Delta or in Wye connection without neutral wire. The same consideration can be made from the perspective of the mitigation control. Thus, single-phase units are best used for the zero- and units without neutral conductor connection – three-phase PV and EV chargers or battery storage – are used to mitigate the negative-sequence. In the following effects of this sequence inter-coupling are investigated and discussed.

Selective negative-sequence mitigation with three-phase units

In this setup flexible single-phase resources like EV1~ chargers or small PV generation units are assumed to be uncontrollable resulting in $\boldsymbol{\mathcal{S}}^{[\nu]}_{\mathrm{EV1}\sim,k} = \overline{\boldsymbol{\mathcal{S}}}^{[\nu]}_{\mathrm{EV1}\sim,k}$. The chargers exchange power with the grid with a given profile and do not contribute to the mitigation. Dispatch-able resources from three-phase units like PV or EV3~ chargers that have no neutral conductor connection are used for the mitigation. The effect on negative- and zero-sequence component is investigated and depicted for $\mathcal{N}_2$ of the low voltage grid.

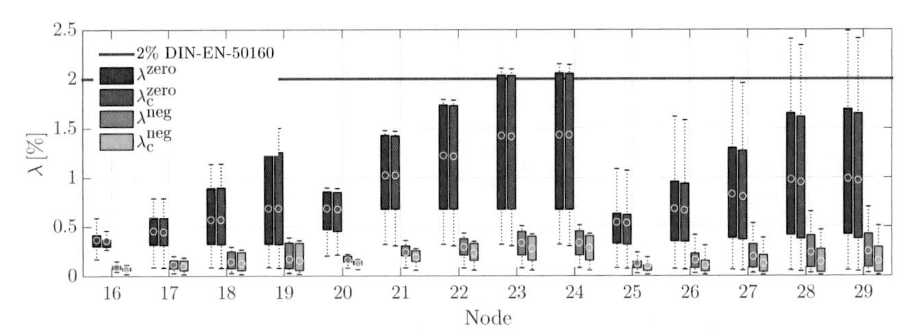

Figure 5.13: Negative- (5.2a) and zero-sequence (5.2b) components from section $\mathcal{N}_2$ in Figure 5.3 in selective negative-sequence mitigation

The mitigation of λ^{neg} is very effective, with a maximum reduction of 1.24 %, as can be seen from the results summarized in Table 5.1. Reducing the negative-sequence also effects λ^{zero} but is insufficient to satisfy the limit of 2 % (red solid line in Figure 5.13.), because (5.10) limits the potential of reducing the neutral current. Thus, it can only be affected by changing the nodal voltage but not through a direct injection of neutral current.

Table 5.1: Absolute reduction of mean, min and maximum unbalance factors in % for each controlled node (selective negative-sequence)

Node	mean $\Delta\lambda^{zero}$	mean $\Delta\lambda^{neg}$	min $\Delta\lambda^{zero}$	min $\Delta\lambda^{neg}$	max $\Delta\lambda^{zero}$	max $\Delta\lambda^{neg}$
5	-0.04	-0.17	0.012	0.003	-0.247	-0.58
8	-0.11	-0.13	0.001	0.015	-1.00	-0.69
15	-0.08	-0.34	-0.004	-0.004	-0.39	-1.24
18	-0.001	-0.018	-0.006	-0.009	0.004	-0.02
21	-0.004	-0.06	-0.015	-0.023	-0.008	-0.08
29	-0.021	-0.09	-0.009	-0.022	-0.064	-0.19
32	-0.005	-0.06	-0.0007	0.01	-0.037	-0.3
34	-0.003	-0.08	-0.0006	0.009	0.0408	-0.32
39	-0.03	-0.21	0.006	-0.02	-0.136	-0.82

Selective zero-sequence mitigation with single-phase units

In this setup flexible three-phase resources like high power EV3∼ chargers or generation units are assumed to be uncontrollable and thus cannot be used by the controller to contribute to the mitigation process. Every power value is equal to the measured node value as $\boldsymbol{\mathcal{S}}^{[\nu]}_{\mathrm{EV3\sim},k} = \overline{\boldsymbol{\mathcal{S}}}^{[\nu]}_{\mathrm{EV3\sim},k}$. Only the single-phase units can be controlled to counteract the unbalance effect of the domestic loads. In Figure 5.14 exemplary result are depicted from section $\mathcal{N}_2$ of the grid. In summary the selective mitigation of λ^{zero}_{ν} also signifi-

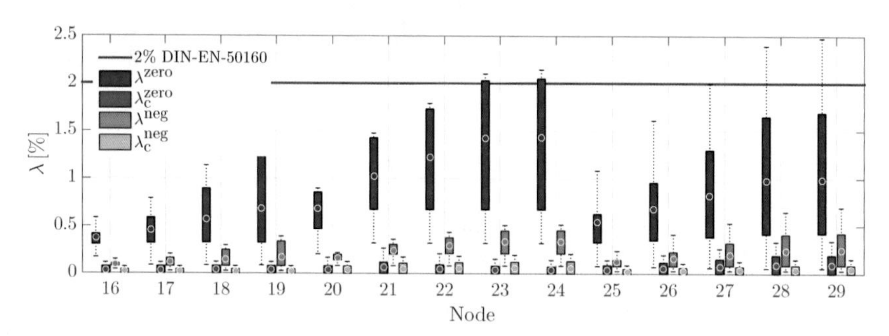

Figure 5.14: Negative- (5.2a) and zero-sequence (5.2b) components from section $\mathcal{N}_2$ in Figure 5.3 with the selective zero-sequence mitigation

cantly reduces λ^{neg}_{ν}. This can be explained because the EV1∼ chargers which are used as actuators reduce the overall single-phase consumption. This not only reduces the

Table 5.2: Absolute reduction of mean, min and maximum unbalance factors in % for each controlled node (selective zero-sequence)

Node	mean $\Delta\lambda^{\text{zero}}$	mean $\Delta\lambda^{\text{neg}}$	min $\Delta\lambda^{\text{zero}}$	min $\Delta\lambda^{\text{neg}}$	max $\Delta\lambda^{\text{zero}}$	max $\Delta\lambda^{\text{neg}}$
5	-0.75	-0.21	-0.061	-0.01	-2.25	-0.65
8	-0.95	-0.28	-0.08	-0.01	-3.5	-1.22
15	-1.47	-0.4	-0.04	-0.003	-4.9	-1.34
18	-0.53	-0.12	-0.08	-0.027	-1.01	-0.21
21	-0.96	-0.19	-0.32	-0.08	-1.21	-0.19
29	-0.88	-0.19	-0.05	-0.03	-2.06	-0.5
32	-0.6	-0.14	-0.08	-0.04	-1.35	-0.3
34	-0.86	-0.20	-0.08	-0.04	-1.65	-0.31
39	-0.98	-0.24	-0.01	-0.02	-3.1	-0.83

neutral but also directly affects the line current and thus the line voltage drops.

Unified sequence-voltage mitigation with all available units

Contrary to the other two scenarios now the negative- and zero-sequence component is mitigated together. The features of each of the locally connected units of the nodes can be used by the controller. Consequently, individual features of each unit can be combined for the different tasks of the mitigation.

The combined strategy further reduces $\lambda_{\nu}^{\text{neg}}$ and $\lambda_{\nu}^{\text{zero}}$. The optimization problem

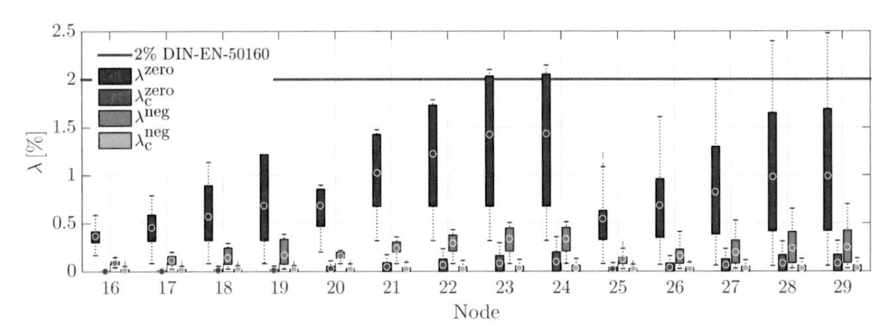

Figure 5.15: Negative- (5.2a) and zero-sequence (5.2b) components from section $\mathcal{N}_2$ in Figure 5.3 in selective negative-sequence mitigation

Table 5.3: Reduction of mean, min and maximum unbalance factors in % for each controlled node (unified)

Node	mean $\Delta\lambda^{\text{zero}}$	mean $\Delta\lambda^{\text{neg}}$	min $\Delta\lambda^{\text{zero}}$	min $\Delta\lambda^{\text{neg}}$	max $\Delta\lambda^{\text{zero}}$	max $\Delta\lambda^{\text{neg}}$
5	-0.79	-0.22	-0.06	-0.012	-2.42	-0.65
8	-0.99	-0.26	-0.07	-0.01	-3.65	-1.11
15	-2.15	-0.6	-0.086	-0.04	-7.19	-2.11
18	-0.56	-0.13	-0.085	-0.03	-1.1	-0.19
21	-0.98	-0.22	-0.33	-0.08	-1.3	-0.25
29	-0.89	-0.2	-0.06	-0.03	-2.1	-0.52
32	-0.57	-0.14	-0.08	-0.03	-1.4	-0.18
34	-0.85	-0.21	-0.08	-0.04	-1.71	-0.25
39	-0.93	-0.11	-0.011	-0.02	-3.21	-0.11

calculates a trade-off of both effects of the unbalance, which can be regulated by the cost term of the local cost function. This strategy has the highest effect on the reduction of the sequence-components with 7.19 % in $\lambda_\nu^{\text{zero}}$ and 2.11 % in λ_ν^{neg}.

5.3.3 Intermediate conclusion

In general, it can be observed, that each unit in the grid, that can provide flexible power either in three- or single-phase, can deliver certain support for voltage balancing. Three-phase inverter units can influence the negative-sequence, but have only limited effect on the zero-sequence, if the neutral conductor is not connected. Single-phase electric vehicle chargers, on the other hand, have a significant effect on the zero-sequence component, because of their direct influence on the neutral conductor current. The combination of both types of flexibility gives the best overall results because the controllers have the possibility to balance the reduction of both components with a certain amount of flexibility in the grid.

Through the cost functions $\mathcal{J}_{\text{glo}}^{[\nu]} + \mathcal{J}_{\text{loc}}^{[\nu]}$ of the local optimization problems, the effect on the sequence-components reduction can be influenced. Additionally, through the weighting $\alpha^{[\nu]}$ of the Jacobi approach, the grid-wide strategy can be adjusted such that the strength of the mitigation participation for each node can be changed.

6 Branch impedance estimation and on-line prediction

Model based control algorithms and optimal power flow need an accurate set of parameters from the underlying power grid to produce reliable results. Especially in low voltage grids with an inherent unbalance in power consumption and limited measurement accuracy from devices like smart-meters, a correct model is even more essential. The following section introduces a distributed neighbor-to-neighbor estimation algorithm based on the approximated power flow developed in Section 2.4.1.

6.1 Introduction and related work

Both control and estimation procedures heavily rely on accurate information of the underlying grid. To acquire this information future low voltage distribution grids have to be equipped with measurement devices and a communication overlay. Currently, the most feasible solution for this data acquisition is a combination of smart-meters and power line communication (PLC) [LTS16]. However, with the current standard in PLC, phase measurement units (PMU) cannot be integrated, because PLC does not have the bandwidth to cope with the needed timing of a PMU [CTV+17]. If 5G wireless technology is available in the next decade, this fact will change in favor for the PMU.

Much of current research has been focused on topology learning of radial distribution grids. An algorithm to estimate the Laplacian and consequently the topology of a radial grid is developed in [BBM+13]. The method uses an approximated power flow model and correlates the voltage magnitude measurements of every node. The assumption that nodal power injections are uncorrelated has to be made which is not entirely correct because of seasonal trends and similar domestic user behavior but is a rather mild assumption. Another approach based on the LinDistFlow [BW89] approximation using only terminal node measurements is developed in [DBC16]. The topology is determined based on a machine learning algorithm. Even though only terminal nodes are measured, hidden nodes and consequently the structure of the grid can be estimated. If the setup is extended with several smart-meters placed at different nodes in the radial network, even simpler approaches with nodal voltage correlation from measurements are possible as presented in [Ben15] and [PGRB16a] as well as [PBPR17]. Even from an information science perspective, the topology estimation can be seen as a very well-established field for radial grids with highly accurate results [ETV13, AL13, PT17]. As a consequence, for this work, it is assumed that the topology of the low voltage grid under investigation is known and only the parameters have to be estimated.

In [JSSM14] a three-phase parameter identification method for distribution grids is developed which assumes measurable branch currents and PMUs. Based on the head and terminal current of the branch and an assumed linear voltage drop across the branch the individual impedances are estimated. However, today and in the recent future, branch-current measurements cannot be assumed in low voltage distribution grids. Another approach that considers measurement and quantization errors, was developed in [PGK+16]. Filtering models and quantization error estimation procedures are deployed. The authors use a transformed version of the nonlinear power flow equations for balanced grids and only use voltage magnitude and apparent power measurements. Another approach which uses the DC power flow approximation for simultaneous estimation of topology and parameters is developed in [LPS13]. The method is computational very demanding and relies on the DC power flow approximation [SJA09] which is not valid in low voltage grids, due to the high R/X ratio. Methods in literature which use simpler approximations for the voltage drop on the branches to determine the grid parameters based on smart-meter data are presented in [PGRB16b, PGRB16a, Pep16]. Although the methodology gives very good results it uses the well known LinDistFlow approximation [BW89] and only works in balanced three-phase or single-phase grids.

The before mention publications always assume a balanced single-phase circuit representation for their investigation. Three-phase unbalanced power grids, however, have received less attention regarding parameter estimation in the research community. In [Ben15] the author builds a power flow model based on machine learning but does not directly estimate the different parameters. The result is a parametrization of a quadratic function in active and reactive power. But for model-based balancing methods as developed in this work, the parameters are very important.
Moreover, the influence of conductor heating on the parameter estimation accuracy in low voltage grids and the effect of voltage and current estimation quality due to the change in line parameters has not received much attention. But this factor will gain significance in the future as the high spot loads in the form of electric vehicles grow in number and thus very high load currents and harmonics will become more frequent [DPDB03, MCR15]. The authors of [BA07] show that the heating of transmission lines has a great impact on state estimation accuracy. Because the R/X ratio in low voltage grids is $>>1$ the effect of conductor heating on the accuracy is even higher than in transmission levels. Usually, the effect of cable heating is studied for the prediction and real-time ampacity rating of cables [Mil06, Deg15]. To the best knowledge of the author, the inclusion of thermal effects into a power flow approximation has not been studied

before.

In a standard setup, the insulation and core temperature of the cable is not measurable [Deg15]. In this work, it is assumed that the soil temperature can be extrapolated or measured and used as an input to the thermal model. The soil, too has a great effect on the admissible ampacity of the cable, because in low voltage grid, they are only buried in 60 cm depth [OO04]. To include this, the LIPF update described in Section. 2.4.2 is not only able to account for local voltage magnitude and power measurements but recalculates the impedance matrix if needed. This way the resistance change due to temperature predictions can be in-cooperated, too.

6.1.1 Preliminaries

Thermal model for low voltage cables

During the operation of the grid, the current flow through the underground cables generate heat by dissipation in the electric resistance. An electric equivalent circuit can be used to model this heat transfer with the losses as a source and the soil in the ground as a sink. Instead of distributed parameters, lumped ones in the form of thermal capacitances and resistances are sufficient to evaluate the heat transfer up to a certain accuracy [Deg15][Mil06]. In the following only heat conduction is considered, as low voltage underground cables are buried in the soil and not surrounded by any cooling liquid.

If no special need for mechanical safety is needed, the majority of German low voltage distribution cables are of type N(A)YY [Yin11]. Four concentric solid aluminum conductors are surrounded by either polyethylene (PE) or polyvinyl-chloride (PVC) conductor insulation (2) and filling materials (3). On the outside of the cable, there is another layer of insulation from PVC (4) grouping the four conductors together as depicted in Figure 6.1. Part (4) is often referred to as the sheath.

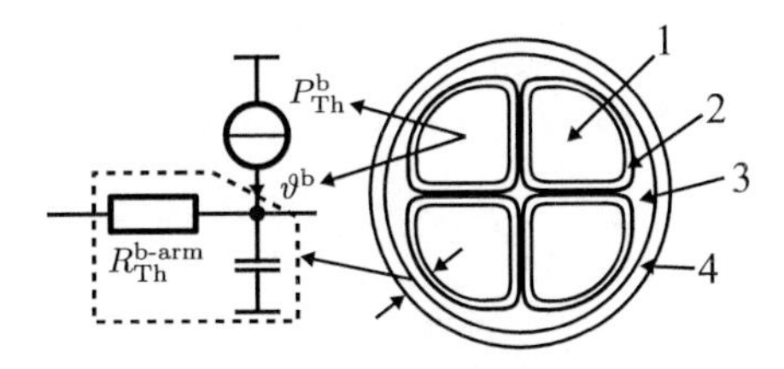

Figure 6.1: Cross sections of a N(A)YY low voltage cable with four conductors on the right and a simple thermal equivalent circuit on the left

In cables, the heat source are the resistive losses which can be described as a function of current flow together with the conductor resistance $\Re\{\underline{z}^{\mathrm{a}}_{\nu,\mu}\} = r^{\mathrm{a}}_{\nu,\mu}$. The thermal power generated in every core of the cable depicted in Figure 6.1 can be calculated as

$$P^{\mathrm{a}}_{\mathrm{Th}_{\nu,\mu}} = \left\|\underline{i}^{\mathrm{a}}_{\nu,\mu}\right\|^2 r^{\mathrm{a}}_{\nu,\mu} \tag{6.1}$$

where $P^{\mathrm{a}}_{\mathrm{Th}}$ is the generated thermal power related to the losses in phase **a** of the cable. The conductor materials such as aluminum and the PVC insulation have specific thermal properties, which can be translated into electric quantities. The reciprocal thermal conductivity, which specifies how well heat can be transported by a medium, can be represented as a thermal resistance R_{Th}. The density and volume of material together with the heat capacity C_{Th} can be transformed into a capacitor. Following the approach in [ERG11] the four-wire low voltage cable from Figure 6.1 can be represented by the thermal equivalent circuit depicted in Figure 6.2.

In contrast to [ERG11] in Figure 6.2 the conductor is modeled as a single heat source

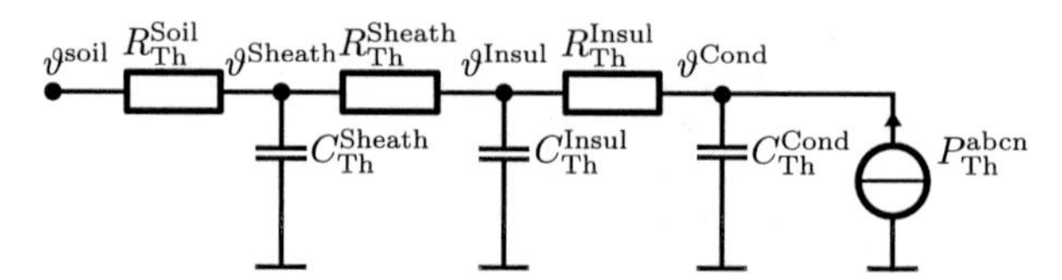

Figure 6.2: Approximated dynamic thermal model for a NAYY $4 \times 120\,\mathrm{mm}^2$ low voltage cable

and not split into the different sources $P^{a}_{\mathrm{Th}} - P^{n}_{\mathrm{Th}}$ for each conductor. They are combined into one element $P^{\mathrm{abcn}}_{\mathrm{Th}}$

$$P^{\mathrm{abcn}}_{\mathrm{Th}} = \left\|\underline{i}^{a}_{\nu,\mu}\right\|^2 r^{\mathrm{a}}_{\nu,\mu} + \left\|\underline{i}^{b}_{\nu,\mu}\right\|^2 r^{\mathrm{b}}_{\nu,\mu} + \left\|\underline{i}^{c}_{\nu,\mu}\right\|^2 r^{\mathrm{c}}_{\nu,\mu} + \left\|\underline{i}^{n}_{\nu,\mu}\right\|^2 r^{\mathrm{n}}_{\nu,\mu} \tag{6.2}$$

for simplicity. The PVC insulation, the extruded filling and the PVC sheath are modeled as cylindrical elements. The equivalent thermal resistance per unit length of these layers can be calculated as [DPDB03]

$$R_{\mathrm{Th}} = \frac{1}{\psi 2\pi} \ln\left(\frac{D_{\mathrm{out}}}{D_{\mathrm{in}}}\right) \tag{6.3}$$

with the inner diameter D_{in} and the outer diameter D_{out} of the cylinder wall and ψ the thermal conductivity. The thermal resistance of the soil is calculated the same way using the buried depth 60 cm of the cable as an equivalent radius.

The materials in the cable have the ability to store thermal energy. This capacity can be calculated by using the material density ρ, the volume V and the specific heat capacity c_{Th} together as

$$C_{\mathrm{Th}} = \rho c_{\mathrm{Th}} \underbrace{\pi \left(\frac{D_{\mathrm{out}}^2}{4} - \frac{D_{\mathrm{out}}^2}{4} \right)}_{V}. \tag{6.4}$$

The change in electrical resistance due to conductor heating can be expressed by a linear function (6.5) if the absolute temperature is below 200 °C.

$$r_{\nu,\mu}^{\mathrm{a}} = r_{\nu,\mu}^{\mathrm{a,20\,C}} \left(1 + \gamma_{\mathrm{AL}} \left(\vartheta^{\mathrm{Cond}} - 20\,^{\circ}\mathrm{C}\right)\right) \tag{6.5}$$

with $r_{\nu,\mu}^{\mathrm{a,20\,^{\circ}C}}$ the reference conductor resistance at 20 °C and $\gamma_{\mathrm{AL}} = 0.004$ the temperature coefficient of Aluminum. The thermal equivalent in Section 6.1.1, Figure 6.2 is used to derive the following thermal state space model for each cable

$$\underbrace{\begin{pmatrix} \dot{\vartheta}^{\mathrm{Cond}} \\ \dot{\vartheta}^{\mathrm{Insul}} \\ \dot{\vartheta}^{\mathrm{Sheath}} \end{pmatrix}}_{\dot{\boldsymbol{\vartheta}}} = \underbrace{\begin{pmatrix} -\frac{1}{R_{\mathrm{Th}}^{\mathrm{Insul}} C_{\mathrm{Th}}^{\mathrm{Cond}}} & \frac{1}{R_{\mathrm{Th}}^{\mathrm{Insul}} C_{\mathrm{Th}}^{\mathrm{Cond}}} & 0 \\ \frac{1}{R_{\mathrm{Th}}^{\mathrm{Insul}} C_{\mathrm{Th}}^{\mathrm{Insul}}} & -\frac{R_{\mathrm{Th}}^{\mathrm{Insul}} + R_{\mathrm{Th}}^{\mathrm{Sheath}}}{R_{\mathrm{Th}}^{\mathrm{Insul}} R_{\mathrm{Th}}^{\mathrm{Sheath}} C_{\mathrm{Th}}^{\mathrm{Insul}}} & \frac{1}{R_{\mathrm{Th}}^{\mathrm{Sheath}} C_{\mathrm{Th}}^{\mathrm{Insul}}} \\ 0 & \frac{1}{R_{\mathrm{Th}}^{\mathrm{Sheath}} C_{\mathrm{Th}}^{\mathrm{Sheath}}} & -\frac{R_{\mathrm{Th}}^{\mathrm{Sheath}} + R_{\mathrm{Th}}^{\mathrm{Soil}}}{R_{\mathrm{Th}}^{\mathrm{Sheath}} R_{\mathrm{Th}}^{\mathrm{Soil}} C_{\mathrm{Th}}^{\mathrm{Sheath}}} \end{pmatrix}}_{\boldsymbol{A}_{\mathrm{Th}}} \underbrace{\begin{pmatrix} \vartheta^{\mathrm{Cond}} \\ \vartheta^{\mathrm{Insul}} \\ \vartheta^{\mathrm{Sheath}} \end{pmatrix}}_{\boldsymbol{\vartheta}} +$$

$$\underbrace{\begin{pmatrix} \frac{1}{C_{\mathrm{Th}}^{\mathrm{Cond}}} & 0 \\ 0 & 0 \\ 0 & \frac{1}{R_{\mathrm{Th}}^{\mathrm{Soil}} C_{\mathrm{Th}}^{\mathrm{Sheath}}} \end{pmatrix}}_{\boldsymbol{B}_{\mathrm{Th}}} \begin{pmatrix} \sum_{\xi=\{\mathrm{a,b,c,n}\}} P_{\mathrm{Th}}^{\xi} \\ \vartheta^{\mathrm{Soil}} \end{pmatrix} \tag{6.6a}$$

$$\tilde{r}_{\mu,\nu}^{\mathrm{pos}} = \begin{pmatrix} \hat{r}_{\mu,\nu}^{\mathrm{pos}} \gamma_{\mathrm{AL}} & 0 & 0 \end{pmatrix} \boldsymbol{\vartheta} + \left(\hat{r}_{\mu,\nu}^{\mathrm{pos}} - \hat{r}_{\mu,\nu}^{\mathrm{pos}} \gamma_{\mathrm{AL}} 20\,^{\circ}\mathrm{C} \right). \tag{6.6b}$$

The input to the dynamic system are the conductor losses in the underground cables $\sum_{\xi=\{\mathrm{a,b,c,n}\}} P_{\mathrm{Th}}^{\xi}$ from (6.2) and the soil temperature $\vartheta^{\mathrm{Soil}}$ in 60 cm depth. Note that a temperature prediction of the soil can be done very efficiently and accurate [Deg15]. Due to this fact the soil temperature is assumed to be known.

6.2 Parameter estimation algorithm

The unbalance in power grids can be significant as has been discussed in Section. 5.3. This introduces difficulties in the estimation of line parameters. On one hand local voltage magnitude measurements are not sufficient anymore to determine the real complex phase voltages, if the unbalance is above a certain threshold (Section. 5.3.1). On the other hand the neutral conductor current introduces an offset in the reference potential that can not be determined by local smart-meters without additional measurement equipment [Kar]. This voltage drop leads to a zero-sequence component that cannot be correctly estimated from the local measurement available (see. Section. 5.3.1 for a discussion on this topic).

Measurement data acquisition

In each node, data points have to be collected for several time instances $k = \{1, \cdots, N_\mathrm{E}\}$. It is important that the amount of samples is higher than the amount of parameters that have to be estimated $N_\mathrm{E} >> N_i$ to receive a good estimation [BCJ+13]. Furthermore, the linear least squares algorithm that is used in this work, needs a solid data base, which makes a preprocessing of the collected measurement necessary. Thus, further only measurements that satisfy the criteria (6.13)-(6.14) are considered for the estimation procedure.

6.2.1 Positive-sequence estimation model

Based on the approximated power flow equations (2.29) it is possible to define a relationship between local measurements of active and reactive power and the voltage magnitude. With $\left\| \left[u_\nu^{\xi,\mathrm{Re}}, u_\nu^{\xi,\mathrm{Im}} \right]^\mathrm{T} \right\|$, $\xi \in$ a,b,c in node $\nu \in \mathcal{N}$ and the injected apparent power $\boldsymbol{\mathcal{S}}$ of every node and phase, the sequence-current injection can be calculated.

To arrive at this point, first the following relationship for the magnitude of the voltage difference is established

$$\left\| \begin{pmatrix} \Delta u_{\nu,k}^{\xi,\mathrm{Re}} \\ \Delta u_{\nu,k}^{\xi,\mathrm{Im}} \end{pmatrix} \right\| = \left\| \mathbb{2}_\nu^{\xi,\mathrm{T}} \mathbb{C}_\mathrm{PF} \left(\mathbb{A}_\mathrm{PF} \right)^{-1} \mathbb{B}_\mathrm{PF} \overline{\boldsymbol{\mathcal{S}}}_k \right\| . \tag{6.7}$$

However, if there is no reactive power in-feed and the voltage is close to nominal, the phase angle difference $\varphi_0^\xi - \varphi_\nu^\xi$, $\xi \in$ a,b,c tends to be small. Based on these assumptions (6.7) can be further simplified to

$$\Delta \overline{u}_{\nu,k}^\mathrm{pos} \approx \begin{pmatrix} \mathbf{1}_\nu^\mathrm{T} & \mathbf{0} \\ \mathbf{0} & \mathbf{0} \end{pmatrix} \left(\mathbb{M}_\mathrm{pos} \mathbb{K}_\mathrm{PF} \mathbb{C}_\mathrm{PF} \left(\mathbb{A}_\mathrm{PF} \right)^{-1} \mathbb{B}_\mathrm{PF} \overline{\boldsymbol{\mathcal{S}}}_k \right) . \tag{6.8}$$

The vector $\mathbf{1}_\nu^\mathrm{T}$ now only maps the effect of injection of apparent power $\overline{\boldsymbol{\mathcal{S}}}$ onto the real part of the positive-sequence voltage.

Note that because only smart meter magnitude measurements are available the imaginary part of the positive-sequence is $\Im\{\underline{\overline{u}}_\nu^\mathrm{pos}\} = 0 \rightarrow \underline{\overline{u}}_\nu^\mathrm{pos} = \overline{u}_\nu^\mathrm{pos}$. With (5.24) only the real part of the positive sequence can be estimated.

Furthermore, in Section 2.4.2 it is assumed that the low voltage cables have symmetrical impedances. Applying this to (6.8) and exploiting the linearity and symmetry of the equation it can be further simplified and reordered leading to

$$\Delta\overline{u}_{\nu,k}^{\mathrm{pos}} \approx \begin{pmatrix} \mathbf{1}_{\nu}^{\mathrm{T}} & 0 \\ \mathbf{0} & \mathbf{0} \end{pmatrix} \left(\boldsymbol{C}_{\mathrm{PF}}\mathbb{M}_{\mathrm{pos}}\mathbb{K}_{\mathrm{PF}}\left(\mathbb{A}_{\mathrm{PF}}\right)^{-1}\mathbb{B}_{\mathrm{PF}}\overline{\boldsymbol{\mathcal{S}}}_{k}\right). \tag{6.9}$$

With the formulation (6.9) it is now possible to map the measured three-phase apparent power $\overline{\boldsymbol{\mathcal{S}}}$ onto the positive-sequence voltage drop of each node. But as has been discussed in Section 2.4.1 this restricts the values valid for the estimation to sections with a lower unbalance factors (5.3a) and (5.3b), but results in a higher overall accuracy of the impedance estimation. The direct estimation of the approximated power flow parameters is not recommendable because they are time-varying with the injected power at each node.

Instead, the power flow approximation is substituted with

$$\left(\mathbb{A}_{\mathrm{PF}}\right)^{-1}\mathbb{B}_{\mathrm{PF}}\overline{\boldsymbol{\mathcal{S}}}_{k} \approx \overline{\mathbb{E}}_{k}\overline{\boldsymbol{\mathcal{S}}}_{k}$$

leading to the following formulation for the vector of positive-sequence voltage drops $\Delta\overline{\boldsymbol{u}}_{k}^{\mathrm{pos}}$ and branch-currents $\overline{\boldsymbol{i}}_{\mathrm{B},k}^{\mathrm{pos}}$

$$\Delta\overline{\boldsymbol{u}}_{k}^{\mathrm{pos}} \approx \boldsymbol{C}_{\mathrm{PF}}\mathbb{M}_{\mathrm{pos}}\mathbb{K}_{\mathrm{PF}}\overline{\mathbb{E}}_{k}\overline{\boldsymbol{\mathcal{S}}}_{k} \tag{6.10a}$$

$$\overline{\boldsymbol{i}}_{\mathrm{B},k}^{\mathrm{pos}} \approx \tilde{\boldsymbol{W}}\mathbb{M}_{\mathrm{pos}}\mathbb{K}_{\mathrm{PF}}\overline{\mathbb{E}}_{k}\overline{\boldsymbol{\mathcal{S}}}_{k} \tag{6.10b}$$

The individual listings in the matrix $\boldsymbol{C}_{\mathrm{PF}}$ are a summary of the different branch impedances. It is not recommended estimating its components directly, because then only the nodal currents and not the summation in the branches are used to estimate the impedance parameters. Instead, the matrix is separated into the different branches of the grid but the positive sequence branch current between two adjacent nodes can directly be included. This also cancels out the slack bus voltage which makes no communication channel to the substation transformer necessary.

The calculation of (6.10a) now reduces to

$$\underbrace{\overline{u}_{\nu,k}^{\mathrm{pos}} - \overline{u}_{\mu,k}^{\mathrm{pos}}}_{\Delta\overline{u}_{\nu,\mu,k}^{\mathrm{pos}}} \approx \left(r_{\nu,\mu}^{\mathrm{pos}} \quad -x_{\nu,\mu}^{\mathrm{pos}}\right) \sum_{\nu>\mu,\nu,\mu\in\mathcal{N}_i} \mathbf{2}_{\nu}^{\mathrm{T}}\tilde{\boldsymbol{W}}\mathbb{M}_{\mathrm{pos}}\mathbb{K}_{\mathrm{PF}}\overline{\mathbb{E}}_{k}\overline{\boldsymbol{\mathcal{S}}}_{k} \tag{6.11}$$

with $\Delta\overline{u}_{\nu,\mu,k}^{\mathrm{pos}}$ being the positive sequence voltage drop across the branch impedance and $r_{\nu,\mu}^{\mathrm{pos}}$, $x_{\nu,\mu}^{\mathrm{pos}}$ the resistance and inductance of the branch between node ν and μ respectively.

6.2.2 Measurement data preprocessing

Positive-sequence current injection measure

The error on the sequence-components increases with the unbalance represented by (5.3a) and (5.3b). Thus, for the parameter-estimation the branch current in all three-phases must be calculated and it's positive-sequence extracted. Because in theory the voltage can be fully balanced while the branch current does only consist of a negative- or zero-sequence component, a measure for the amount of suitable current flow over the branches is defined, by comparing the amount of unbalance related current components to the geometric mean. The relationship can be formulated as

$$\zeta_\nu = \frac{\left\| \mathbf{2}_\nu^T \tilde{\boldsymbol{W}} \mathbb{M}_{\text{neg}} \mathbb{K}_{\text{PF}} \overline{\mathbb{E}}_k \overline{\boldsymbol{\mathcal{S}}}_k \right\| + \left\| \mathbf{2}_\nu^T \tilde{\boldsymbol{W}} \mathbb{M}_{\text{zero}} \mathbb{K}_{\text{PF}} \overline{\mathbb{E}}_k \overline{\boldsymbol{\mathcal{S}}}_k \right\|}{\left\| \mathbf{2}_\nu^T \tilde{\boldsymbol{W}} \overline{\mathbb{E}}_k \overline{\boldsymbol{\mathcal{S}}}_k \right\|}. \tag{6.12}$$

The current injection unbalance factor ζ_ν from (6.12) takes on values in the range of $\{0, \cdots, 1\}$ where 1 is completely unbalanced and 0 equals fully balanced.

If the grid section that has to be estimated has a relatively low impedance the effect of the components in the denominator of (6.12) can have nearly no effect. Moreover, the unbalance factors (5.3a) and (5.3b) are not useful because as investigated in Section 5.3.1 zero- and negative-sequence are equal and underestimate the true unbalance. That is why for the voltage another factor has to be implemented to judge if the values can be processed for the parameter estimation.

Voltage unbalance estimation based on geometric distance

The unbalance definition from [PM01] based on ANSI/IEEE 112-1991 is used here as a secondary indicator. It should be directly pointed out that this value cannot be used for mitigation, but rather gives an overall unbalance value based on the voltage magnitude. It is used here because it gives a more accurate result if only magnitude measurements are available.

$$\delta_\nu = \frac{\max\{\|\underline{\overline{u}}_\nu^{\text{a}}\|, \|\underline{\overline{u}}_\nu^{\text{b}}\|, \|\underline{\overline{u}}_\nu^{\text{c}}\|\} - \frac{1}{3}\left(\|\underline{\overline{u}}_\nu^{\text{a}}\| + \|\underline{\overline{u}}_\nu^{\text{b}}\| + \|\underline{\overline{u}}_\nu^{\text{c}}\|\right)}{\frac{1}{3}\left(\|\underline{\overline{u}}_\nu^{\text{a}}\| + \|\underline{\overline{u}}_\nu^{\text{b}}\| + \|\underline{\overline{u}}_\nu^{\text{c}}\|\right)} \tag{6.13}$$

The indicator uses the maximum value from the voltage magnitudes of all phases, subtracts the mean value and normalizes it with the mean value. It is in the range of $0 \leq \delta_\nu \leq 0.2$, with 0.2 being the maximum distance from the nominal value. It should be noted, that the value of δ_ν (6.13) can be in range of 0.1-0.2 even though the local power injection of the node is balanced.

Positive-sequence voltage drop

Only the real part of the positive-sequence voltage is considered in the estimation model (6.11). But the deviation of the positive-sequence from its nominal value can have an influence on the voltage angle and thus on the imaginary part, which is neglected. An angle change in low voltage grids is always coupled with a change in magnitude and vice versa. The reason for this can again be found in the low X/R ratio. By keeping the voltage deviation of the mapped measurements in a certain range around the nominal value, the quality of the estimation can be further increased.

$$\psi_\nu = \left\| \overline{u}^{\text{pos}}_{\nu,k} - u_0 \right\| \tag{6.14}$$

Measurement values for the impedance estimation are only considered if the indicators ζ_ν,δ_ν and ψ_ν are below some threshold, otherwise they are not used.

6.2.3 Positive-sequence impedance estimation

The nodal voltage difference between nodes $\Delta u^{\text{pos}}_{\nu,\mu,k}$ can be determined based on the collected three-phase measurements and calculated by first mapping the voltage magnitudes with (5.24) and then transforming them with (2.28). The active and reactive power values are also exchanged by the controller overlay, over the communication bus. The exchange of information is only necessary between two adjacent nodes. The reason for this is, that the branch current in radial networks is always the sum of currents of the connected nodes. So starting from a terminal node, each pair of neighbors gains knowledge of an additional part of the overall branch current as depicted in Figure 6.3.

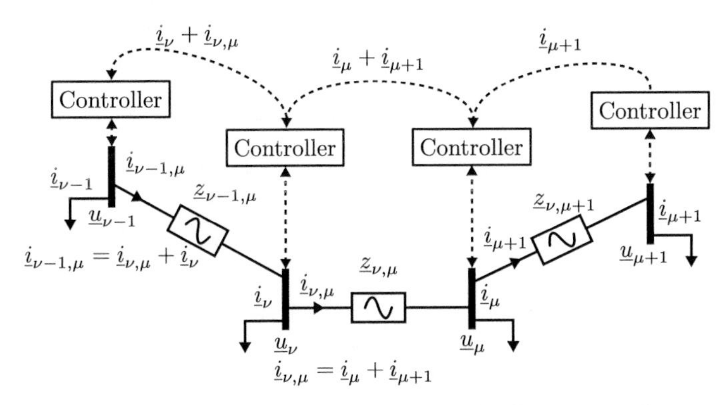

Figure 6.3: Communication infrastructure and principal information exchange setup for the branch impedance estimation procedure

Linear least squares estimation

The least squares problem to estimate line impedance parameters boils down to

$$\underbrace{\begin{pmatrix} \Delta u^{\mathrm{pos}}_{\nu,\mu,k} \\ \Delta u^{\mathrm{pos}}_{\nu,\mu,k+1} \\ \vdots \\ \Delta u^{\mathrm{pos}}_{\nu,\mu,k+N_{\mathrm{E}}} \end{pmatrix}}_{\Delta \boldsymbol{u}^{\mathrm{pos}}_{\nu,\mu}} = \underbrace{\begin{pmatrix} \mathbf{2}^{\mathrm{T}}_{\nu}\tilde{\boldsymbol{W}}\mathbb{M}_{\mathrm{pos}}\mathbb{K}_{\mathrm{PF}}\overline{\mathbb{E}}_{k}\,\overline{\boldsymbol{\mathcal{S}}}_{k} \\ \mathbf{2}^{\mathrm{T}}_{\nu}\tilde{\boldsymbol{W}}\mathbb{M}_{\mathrm{pos}}\mathbb{K}_{\mathrm{PF}}\overline{\mathbb{E}}_{k+1}\,\overline{\boldsymbol{\mathcal{S}}}_{k+1} \\ \vdots \\ \mathbf{2}^{\mathrm{T}}_{\nu}\tilde{\boldsymbol{W}}\mathbb{M}_{\mathrm{pos}}\mathbb{K}_{\mathrm{PF}}\overline{\mathbb{E}}_{k+N_{\mathrm{E}}}\,\overline{\boldsymbol{\mathcal{S}}}_{k+N_{\mathrm{E}}} \end{pmatrix}}_{\mathbb{I}} \underbrace{\begin{pmatrix} r^{\mathrm{pos}}_{\nu,\mu} \\ x^{\mathrm{pos}}_{\nu,\mu} \end{pmatrix}}_{\boldsymbol{z}^{\mathrm{pos}}_{\nu,\mu}} \tag{6.15}$$

To receive the parameters r^{pos}_{ν}, x^{pos}_{ν} the goal is to minimize the difference

$$\Delta \boldsymbol{u}^{\mathrm{pos}}_{\nu,\mu} - \mathbb{I}\boldsymbol{z}^{\mathrm{pos}}_{\nu,\mu}$$

By solving the following unconstrained optimization problem

$$\min_{r^{\mathrm{pos}}_{\nu},\, x^{\mathrm{pos}}_{\nu}} \quad \left(\Delta \boldsymbol{u}^{\mathrm{pos}}_{\nu,\mu} - \mathbb{I}\boldsymbol{z}^{\mathrm{pos}}_{\nu,\mu}\right)^{\mathrm{T}} \left(\Delta \boldsymbol{u}^{\mathrm{pos}}_{\nu,\mu} - \mathbb{I}\boldsymbol{z}^{\mathrm{pos}}_{\nu,\mu}\right) \tag{6.16}$$

As a well known fact (6.16) has an analytic solution [BV04] in the form of

$$\hat{\boldsymbol{z}}^{\mathrm{pos}}_{\nu,\mu} = \begin{pmatrix} \hat{r}^{\mathrm{pos}}_{\nu,\mu} \\ \hat{x}^{\mathrm{pos}}_{\nu,\mu} \end{pmatrix} = \left(\mathbb{I}^{\mathrm{T}}\mathbb{I}\right)^{-1}\mathbb{I}^{\mathrm{T}}\Delta \boldsymbol{u}^{\mathrm{pos}}_{\nu,\mu} \tag{6.17}$$

Based on the estimated positive-sequence impedance of every branch the resistance and inductance vector $\hat{\boldsymbol{r}}^{\mathrm{pos}}_{\mathrm{B}}$, $\hat{\boldsymbol{x}}^{\mathrm{pos}}_{\mathrm{B}}$ is received by stacking up all values in chronological order.

Note that the solution of (6.17) only estimates the positive-sequence component parameters that are related to the real part of the positive-sequence nodal voltage. However, because the real-valued decomposition of complex multiplications leads to skew-symmetric matrices as shown in (2.31), there is no need to estimate the imaginary part directly. The real part of the positive-sequence voltage is also effected by the imaginary part of the impedance and if its influence is high enough, it can be recast from the real part of the positive-sequence as well.

Furthermore, the least-squares algorithm in (6.17) is sensitive to outliers, because it aims to minimize the sum of squares. So, in other words, the squared distances from the estimated values to the actual data-points is minimized. In the case of outliers, the algorithm tries to achieve this even for individual extreme data-points [Sim06, Pep16]. So for better performance, a bad data detection algorithm as in [AGM16], would have to be applied for a real setup, too.

6.2.4 Thermal model to track on-line parameter variation

With the estimated cable parameters the model (6.6b) is used to determine the resistance change of the cable and use this information to update the LIPF. In return the interaction of the different parts of this work are depicted in Figure 6.4

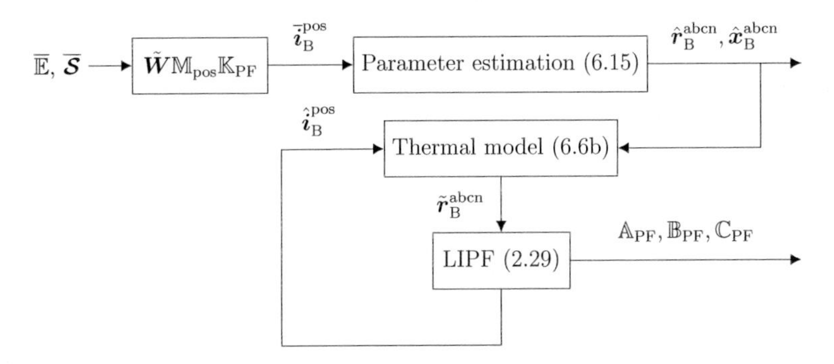

Figure 6.4: Dependency graph between the parameter estimation the thermal cable model and the approximated power flow

This model is used to track the parameter variation, due to the thermal loading of the cables. Figure 6.2.4 summarizes the algorithm with it's different parts in a flowchart.

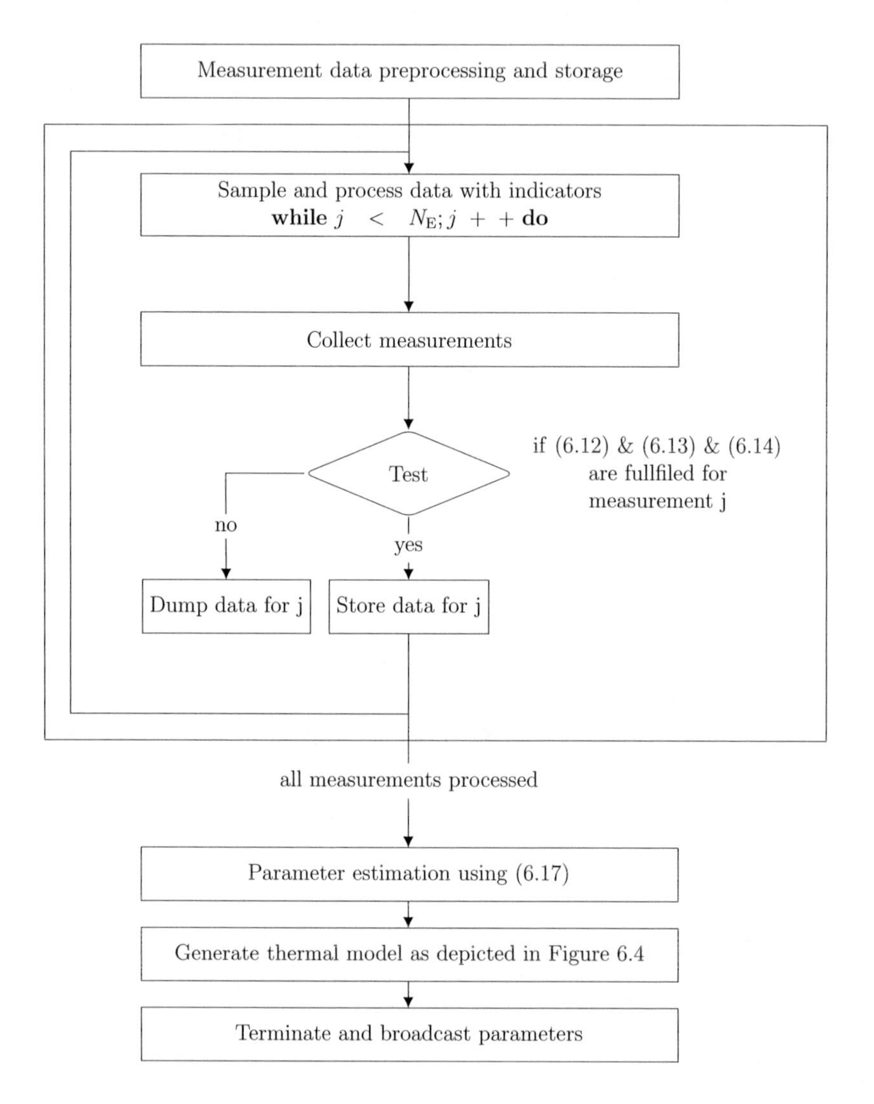

Figure 6.5: Algorithm flowchart depicting the steps of the parameter estimation algorithm in compact form

6.3 Evaluation of the estimation strategy based on simulation

To investigate the effectiveness of the proposed algorithm, a simulation study is conducted. The full parameter values for the cables used in the investigation can be found in appendix B.

The investigation is split into two parts:

Mean branch impedance estimation: in the first part, the mean impedance of the branches is evaluated. For this, the power injections and nodal voltages resulting from the unified balancing approach in Section 5.3.2 are utilized. The magnitudes of the complex voltages are calculated from these results to function as virtual smart-meter measurements.
Electrical resistance changes of the aluminum conductor due to the cable heating by the branch currents are implemented in the underlying power grid model, but treated as unavailable for the estimation. Through this first part of the procedure, a basic knowledge of the impedance of the branches is gained.

Variable impedance prediction for voltage and current estimation: secondly, the variability of the resistive part of the cable impedance is considered based on the estimated parameters. By deploying the thermal cable model, the core temperature of the underground cables is simulated. This in turn gives the possibility to monitor the thermal change of the cable. Thus, an actual resistance value is available, without the need to run the estimation process for small-time windows. The variation in resistance is applied to the approximated power flow according to Figure 6.4 and compared with OpenDSS using true impedance parameters. The effect on the voltage and branch current mismatch is then investigated.

6.3.1 Single-shot estimation for mean impedance

It is assumed that local measurements are available for one day, but with a high granularity of 1 s. The single shot estimation gives some basic knowledge about the grid impedances for the individual controllers. For that purpose the branch impedances of the 40 node benchmark grid are estimated based on (6.15) and the accuracy of the estimation procedure is investigated for the use case that the balancing controllers of Section 5.2 are deployed on the benchmark circuit and power injections as well as voltage magnitude measurements from smart-meters are collected after the mitigation process is completed. This way the circuit is much more balanced and thus the accuracy of the estimation is higher.

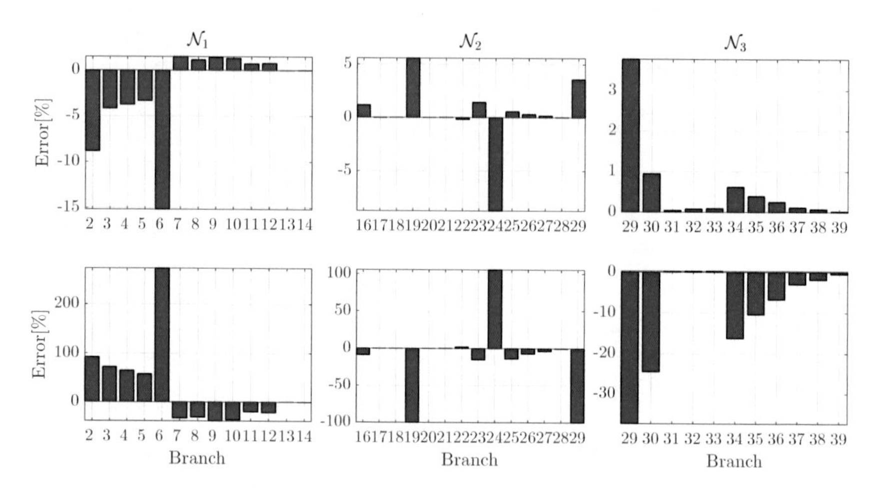

Figure 6.6: Mean impedance estimation error for cable resistance (upper figures) and inductance (lower figures), for all grid sections $\mathcal{N}_i$, $i \in \{1, 2, 3\}$

The inductance value of N(A)YY 0.6/1 kV low voltage cables is very small [OO04] and thus the effect on the real part of the positive-sequence component is limited. This is reflected in the estimation accuracy depicted in Figure 6.6. The estimation error for the resistance and inductance is listed for the case of unknown cable types in Table 6.1

Table 6.1: Branch impedance estimation accuracy / no cable type knowledge

Section	mean $\frac{r_{\text{real}}-r_{\text{est}}}{r_{\text{real}}}$ [%]	max $\frac{r_{\text{real}}-r_{\text{est}}}{r_{\text{real}}}$ [%]	mean $\frac{x_{\text{real}}-x_{\text{est}}}{x_{\text{real}}}$ [%]	max $\frac{x_{\text{real}}-x_{\text{est}}}{x_{\text{real}}}$ [%]
$\mathcal{N}_1$	-2.18	-15.1	29.26	273.2
$\mathcal{N}_2$	0.28	-8.7	-10.1	105.6
$\mathcal{N}_3$	0.58	3.77	-9.08	-37.3

Impedance matrix construction

If the type of the underground cables is known, the accuracy of the result for the estimated inductance can be increased. From the resistance and the cable type, the length and consequently the equivalent inductance can be determined. Usually the cable type is available from the database of a DSO. This way the overall estimation error can be reduced significantly. Based on the known cable types and thus known X/R ratio, the inductance values are calculated based on the estimated resistance. From this branch impedance the balanced matrix (2.1) and in the next step the three-phase four-wire representation (2.3) + (2.5) can be constructed.

6.3.2 Nodal voltage and branch-current prediction with thermal model

The low voltage cable resistance can change up to 20 % while heating up from the soil temperature to maximum operation temperature of 70°C. The initial parameter estimation from Section 6.3.1 only catches a mean value of the branch impedance parameters, because the acquired data sets are usually for different operating points during the day. With the help of (6.6b) the electro-thermal coupling in the cables is represented based on a simple dynamic model. The cable losses, which are an input to this model are generated with the predicted branch currents from the power flow approximation. By updating the approximation based on the scheme depicted in Figure 6.4 the variation in impedance parameters are directly included. Both results with and without updating the parameters from the base scenario (mean value estimation) are discussed in the following. The 40 node benchmark grid and the same parameter set as for the mean impedance estimation is used for the prediction.

Voltage magnitude prediction error

First the voltage magnitude calculation of the power flow approximation with the parameters from the estimation is compared with the results of OpenDSS.
Especially in the branch of nodes 26-29 depicted in Figure 6.7, the voltage error $\Delta \left\| \underline{u}_{\nu}^{\xi} \right\|$ is high. The reason for this is that the these nodes are loaded heavily with unbalanced consumers like single-phase electric vehicle chargers. A second reason is that the impedance error in the line and the neutral conductor for the fixed impedance model are added, resulting in a strong mismatch in the calculated voltage drop. Furthermore, the maximum impedance error does not have a significant influence on the voltage magnitude estimation accuracy. This can be explained by the fact that the highest impedance estimation error is located at the branch with the lowest impedance value. The voltage drop across this particular branch is minimal and can be neglected.

Through the inclusion of the dynamic model (6.6b), the maximum error in voltage magnitude prediction is reduced from nearly one percent to 0.2 %. The estimation results are depicted in Figure 6.8. Considering the voltage magnitude bounds of ±10% the voltage prediction accuracy is sufficient for the control algorithm and summarized

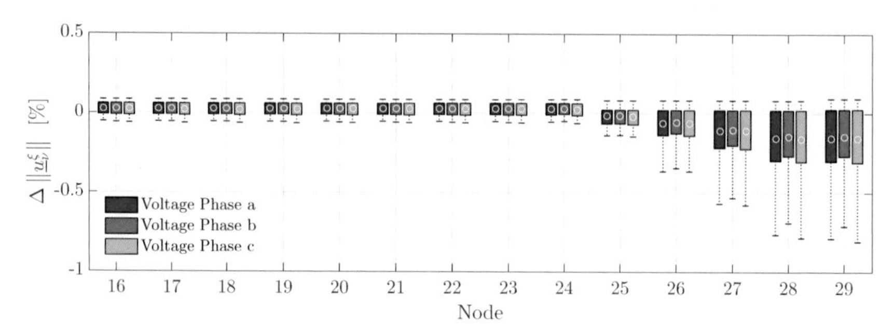

Figure 6.7: Voltage and branch current magnitude error for from section $\mathcal{N}_2$, neglecting thermal effects

in Table 6.2. Another error which is important for the control is the branch current

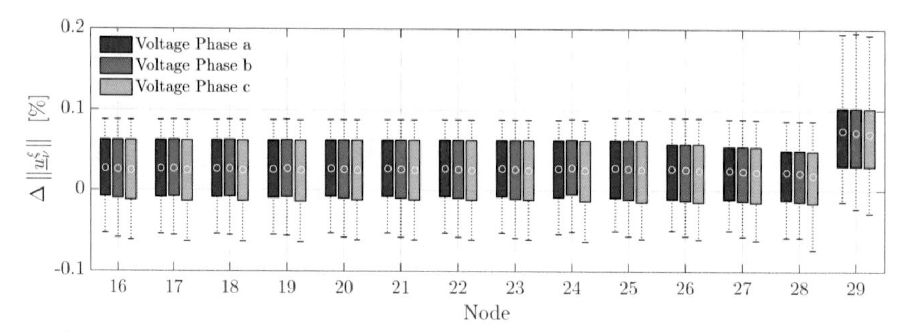

Figure 6.8: Voltage and branch current magnitude error for from section $\mathcal{N}_2$. On-line prediction with the thermal cable model

error. The higher the accuracy of the branch current estimation, the lower the risk of an overloading.

Branch-current magnitude error

Again the branches that have the highest loading suffer from the greatest estimation error $\Delta \left\| \underline{i}_{\mathrm{B},\nu}^{\xi} \right\|$ as depicted in Figure. 6.9 for the case of neglecting any thermal effects. In some branches, especially the ones with low current (branch 17-24), the heating of

Table 6.2: Node voltage magnitude estimation error for the assumption with constant and real time thermal model (Th)

Section	max error phase a [%]	max error phase b [%]	max error phase c [%]
$\mathcal{N}_1$	0.593	0.551	0.561
$\mathcal{N}_1$(Th)	0.147	0.098	0.113
$\mathcal{N}_2$	0.91	0.9	0.91
$\mathcal{N}_2$(Th)	0.194	0.196	0.195
$\mathcal{N}_3$	0.58	0.6	0.58
$\mathcal{N}_3$(Th)	0.112	0.113	0.111

the conductor has only a very small effect on the current accuracy. Nevertheless, the

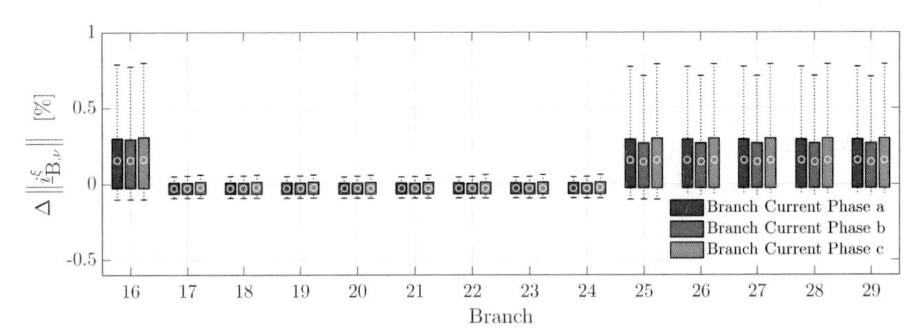

Figure 6.9: Branch current magnitude error for section $\mathcal{N}_2$ neglecting any thermal effects

results of the prediction based on the thermal model (Figure 6.10) show a tendency to increase the accuracy of the linear interpolation power flow. It is important to point out that this accuracy gain is proportional to the loading of the line. This, in return, means that the more the capacity of the cables is used, the more the control system benefits from the thermal model. In Table 6.3 the result of the comparison are summarized for all three sections of the grid, which have been depicted in Figure 5.3.

Intermediate conclusion

The one-shot estimation accuracy is sufficient for the application but will be even less accurate for a practical-setup. Either, a longer period e.g. several weeks of data must be used for the one-shot estimation, or there must be further adjustments in the algorithm.

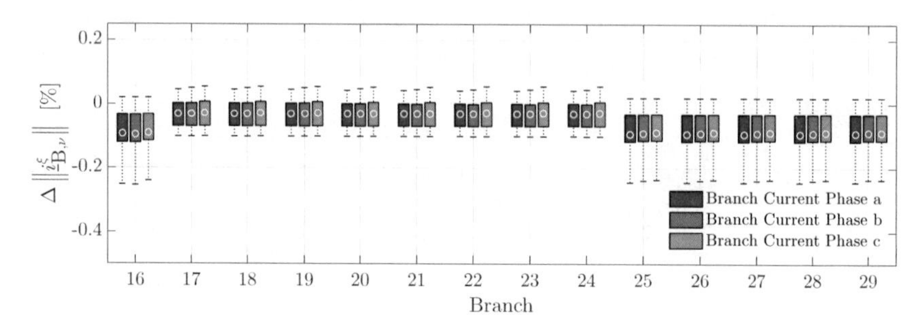

Figure 6.10: Branch current magnitude error for section $\mathcal{N}_2$. Real-time prediction with thermal model

Table 6.3: Branch current estimation error for the assumption of constant impedances and real-time thermal model (Th)

Section	max error phase a [%]	max error phase b [%]	max error phase c [%]
$\mathcal{N}_1$	0.7	0.64	0.73
$\mathcal{N}_1$(Th)	0.639	0.423	0.284
$\mathcal{N}_2$	0.94	0.93	0.94
$\mathcal{N}_2$(Th)	0.305	0.306	0.305
$\mathcal{N}_3$	0.58	0.57	0.58
$\mathcal{N}_3$(Th)	0.153	0.152	0.156

The inductance estimation, on the other hand, can only be achieved with satisfactory estimation quality if the cable type is known. The inductance value of low voltage cables of several meters is always in the range of $\mu\Omega$.
The inclusion of the dynamic thermal-model and the update functionality shows good results. The effectiveness of the thermal model depends on the conductor cross-section and the length of the cable. Thus, a stronger change in cable resistance increases the positive effect that the proposed approach has on the voltage and current prediction. In summary, the following results have been achieved:

- Unbalance indicators were developed, that can be used to filter measurements during a high unbalance in the grid. Simple magnitude measurements from smart-meters are sufficient, even though the grid is unbalanced.
- A linear least-squares algorithm was introduced that estimates branch impedances. The formulation maps the three-phase apparent power values onto the positive sequence-impedance values.

- Through a simplified dynamic thermal model of the cable, the change in impedance value is tracked on-line. Thus, the prediction quality of the power flow approximation is further increased through the feedback of the thermal model.

7 Conclusions and outlook

7.1 Conclusion

The immense growth of distributed generation units and flexible demand introduces the need for active distribution system operation. Especially in low voltage grids with an increase of photo-voltaic units and electric vehicles, a worst case network extension is no longer economically feasible. Instead ancillary services based on mathematical optimization and their beneficial properties have been introduced in power system related research. If these methods are combined with distributed control mechanisms the resulting distributed model predictive controllers are able to handle complex structures, hard system limitations and make it simple to integrate and pursue economical goals. In this thesis, a distributed control framework for an optimized operation of low voltage grids was developed. Essential challenges of future distribution grids are approached by the combination of a novel modular and explicit power flow approximation with distributed control, to solve specific problems in low voltage power system operation.

Explicit linear power flow approximation

A novel power flow approximation is developed in **Chapter 2**, which is based on the parameterization of a linear matrix equation. The difference to classical approaches is that it is flexible with regard to an operating point of the grid i.e. not limited to an expansion with respect to a no load solution. Instead the approximation is parameterized with an on-line feedback from grid measurements. This offers the flexibility to adapt the formulation with acquired data from phase measurement units or smart-meters, increasing the overall accuracy of the solution. The comparison to state of the art power flow approximation approaches from literature, shows that the performance is superior especially when the curvature of the voltage profile is increased. This in turn means that the border areas of admissible nodal voltages can be calculated with a good accuracy. Different simulation studies showed that voltage magnitude measurements from smart-meters are sufficient to receive a good power flow approximation quality, which is not the case for other methods from literature.

Distributed control strategy for operation management

In **Chapter 4** a distributed controller for the optimal dispatch of balanced three-phase renewable generation and demand management is developed. Local controllers in the grid jointly agree on the solution to a dynamic optimal power flow problem, that is integrated in a distributed model predictive control framework. Instead of communicating power trajectories the controllers exchange constraints in the form of performance charts, to decide which dispatch solution for the overall grid is most economical. This way local information about the consumer is transformed and kept private. Each controller has a global view on the power flow in the grid and as such the distributed controller setup is resilient against controller failure.

The underlying equations of the dynamic optimal power flow problem are efficiently solved with the on-line adaptable approximated power flow developed in **Chapter 2**. It is shown that the linear power flow can exploit the receding-horizon scheme of the proposed distributed control scheme by using the locally available measurements of apparent power and voltage magnitude. The result of different simulation studies show that the distributed control enforces branch current and voltage limitation in the grid and at the same time maximizes the benefit for local nodes. In the scenarios customer profit is preferred charging of electric vehicles and the active power in-feed of their own photo-voltaic plants. At the same time locally installed battery storages are only used by the controllers to provide services for peak- and load shaving as well as reactive power support.

In order to investigate the feasibility for an implementation, the distributed control as well as a series of benchmark grids are implemented on a dSpace© rapid prototyping platform, which is coupled to a scaled medium voltage grid simulator. This simulator was fitted with additional hard- and software components, as described in **Chapter 3** to represent a smart-grid. The conducted investigations prove the practical feasibility of the distributed control approach and show the advantages of a flexible approximated power flow.

Distributed strategy for unbalance mitigation

Single phase consumers and small photo-voltaic units introduce a load unbalance in the low voltage distribution grid. To overcome this a distributed algorithm is developed in **Chapter 5** to mitigate this effect. The coupling between the strategy for operation management with distributed model predictive control and the unbalance mitigation strategy allow for simpler algorithms to be used, because system limitations are already taken care of. Flexible resources from different inverter units which are not occupied by the overlying dynamic optimal power flow are locally available. The linear power flow approximation again plays a vital role for the simplification of a three-phase optimal power flow, which is formulated locally for every node. This is achieved by a suitable decomposition, which allows for the reformulation of the different symmetrical components of voltage, as a local function of apparent power injections. After the decomposition only small local optimization problems need to be solved in order to mitigate the unbalance for the whole power network. To further increase the effectiveness of the mitigation a coordination of the individual solutions is implemented with a Jacobi-like algorithm that is applied to find the best power allocation.

In different simulation studies the effectiveness of the proposed algorithm is demonstrated. A selective mitigation strategy that only reduces negative- or zero-sequence voltage components and a unified approach that mitigates both simultaneously are investigated. The results show that the zero-sequence in low voltage grids is the main driving force of the unbalance and thus the selective zero-sequence and the unified approach are the most effective. Due to this fact, the dominant unbalance is best mitigated with single-phase units.

Parameter estimation strategy

Model based control algorithms and optimal power flow need an accurate set of parameters from the underlying power grid to produce reliable results. In **Chapter 6** an algorithm to estimate the cable parameters of the grid is developed. The approach can directly estimate the positive-sequence parameters of the branch impedance from the unbalanced three-phase magnitude measurements of smart-meters. Furthermore, the inclusion of a dynamic thermal model tracks the parameter variation due to a current flow in the conductors of the cable. The temperature change by ohmic dissipation in the cables is modeled by a first-order dynamic equation with the assumption of measurable soil temperature.

To exclude unwanted effects of the unbalanced voltages, indicators where developed to pre-process the measurements in a first step. This is necessary when only voltage magnitude measurement available because the accuracy of positive-sequence voltage estimation deteriorates with increasing unbalance. The unbalance in the branch current is calculated and an indicator is used to filter only values below a certain threshold. After processing the measurements the mean impedances of the cables are estimated with a recursive least-squares algorithm. With some additional mild assumptions the impedance matrix of the grid can then be constructed with the estimated parameters based on known topology information which is assumed to be available from the distributed system operator.
After the grid impedance matrix is constructed a second problem is approached. The thermal behavior of the cables is included in the investigated benchmark grid with the developed dynamic model. A combined model is then used to characterize the voltage and current calculation accuracy based on the approximated power flow. Based on the

achieved results it can be concluded that the parameter estimation for the inductance value can only be done sufficiently if either the cable type is known, or the cable cross section is small i.e. the cable inductance can be neglected.

7.2 Outlook

Explicit linear power flow approximation

Currently this method does only support radial grids without meshes. But nevertheless, it would be interesting to investigate how well it would perform in case of light meshing. Since it is derived from the general Forward-Backward-Sweep method, for which a known extension for weakly meshed grids is available [ADF+10]. This would further increase the range of possible applications, because especially grids in urban areas and cities have a meshed character.

Distributed control strategy for operation management

The distributed operation strategy presented in **Chapter 4** currently only has one negotiation cycle implemented. This has been done to drastically reduce the communication effort for a real-time implementation. However, it could be very interesting to investigated additional negotiation cycles or even other approaches that iterate till a convergence is reached.
A second point which is worth investigating is the use of the thermal cable model for transport capacity prediction. It has been shown to be very effective to predict cable ampacity [Deg15], but has to the best knowledge of the author, not been investigated in combination with model predictive control. Including the thermal model in the distributed control could result in an even more efficient use of the power system capacity.

Distributed strategy for unbalance mitigation

So far it was assumed that a reduction in single-phase charging power, which is used to balance the circuit, does not effect the usability of the electric vehicle. However, in general this cannot be guaranteed, because the optimization has no incentive to fully charge the car. Without this incentive included there is no obligation to reach a certain charging state or increase the length of the scheduled time slot. A user model and better cost function to include this effect would be a good extension.
Another factor is the convergence speed of the Jacobi method. By using more advanced distributed algorithms as in [KZGB16], the overall convergence time could be reduced. However, this factor must always be weighted against the amount of communication that is needed.

Parameter estimation strategy

Further investigation concerning the estimation accuracy of the inductance of the cable should be pursued. This is an important factor, because the inductance value is very small and thus prone to errors.

8 Summary in German

Einführung und Übersicht

Das immense Wachstum von dezentralen Erzeugungseinheiten und das zukünftige Aufkommen von flexiblen Lasten erfordern einen aktiven Betrieb der Verteilnetze. Besonders in Niederspannungsnetzen, in denen die Durchdringung mit Photovoltaikanlagen und Elektrofahrzeugen langfristig sehr hoch sein wird, ist eine Netzkapazitätserweiterung für den "Worst-Case" nicht mehr wirtschaftlich abbildbar. In dieser Arbeit werden alternative Strategien entwickelt, welche netzdienliche Eigenschaften mit dem regulären Betrieb des Netzes kombinieren. Hierzu werden Regelungsverfahren eingeführt, die basierend auf einer mathematischen Optimierung die einzelnen Netzteilnehmer koordinieren und den Leistungsfluss im Netz kontrollieren.
Die entwickelte Betriebsstrategie bedient sich einer Kombination von verteilten modellprädiktiven Regelungen und dynamischen optimalen Lastflussverfahren. Durch die Eigenschaften der beiden Verfahren ist es möglich, komplexe Strukturen, sowie harte Systemgrenzen aktiv in die Regelung zu integrieren und gleichzeitig systemdienliche Ziele zu verfolgen. Es wird gezeigt, dass mit dieser Methode wesentliche Herausforderungen für spezifische Probleme im Niederspannungsnetz gelöst werden. Eine weitere Besonderheit stellt eine neuartige modulare und explizite Approximation dar, welche genutzt wird, um die Lösung des optimalen Lastflusses zu vereinfachen.

Inhaltliche Zusammenfassung der Arbeit

In **Kapitel 2** wird ein neuartiges approximiertes Lastflussverfahren entwickelt, welches genutzt wird, um eine verkürzte Berechnungszeit von modellprädiktiven Regelungsverfahren zu erreichen. Im Gegensatz zu den in der Literatur bekannten Verfahren zeichnet dieses sich dadurch aus, dass es den gleitenden Horizont der modellprädiktiven Regelung geschickt ausnutzt, um für jeden Arbeitspunkt angepasste genaue Ergebnisse zu liefern. Dies wird erreicht, indem die Nichtlinearität eines Konstantleistungs-Modells durch eine Arbeitspunkt abhängige Funktion beschrieben wird. Jedoch wird auf die Nutzung einer klassischen Approximation wie z.B. durch eine Taylor-Entwicklung verzichtet und stattdessen eine lineare Funktion mit variablen Parametern verwendet. Diese Parameter ergeben sich aus den gemessenen Werten von Scheinleistung und Spannung. Es wird in der Arbeit gezeigt, dass die gemessene Spannungsamplitude ausreicht, um genügend genaue Ergebnisse im Vergleich zur nicht-linearen Lösung des Lastflussproblems zu erreichen. Die Konsequenz, die sich hieraus ergibt, ist, dass Smart-Meter im Versorgungsnetz genügen, um die Messwertaufnahme vorzunehmen.

Kapitel 3 stellt die Beispielnetze vor, welche genutzt werden, um die Effektivität der entwickelten Algorithmen zu untersuchen. Die eingeführten Vergleichsnetze sind zum einen ein minimales akademisches Beispiel ausgeführt als vier Knoten Netz, das CIGRÉ Niederspannungsnetz für ländliche und vorstädtische Strukturen und zum anderen ein 40 Knoten Netz, welches aus realen Planungsdaten gewonnen wurde. Des Weiteren wird der praktische Demonstrator mit den wichtigsten Komponenten vorgestellt. Dieser wird zusammen mit einem dSpace System verwendet, um die praktische Realisierbarkeit der Betriebsstrategie zu untersuchen.

Das in **Kapitel 4** entwickelte verteilte Regelungsverfahren beschäftigt sich mit dem Betrieb von zukünftigen Niederspannungsnetzen. Die Regler sind lokal den jeweiligen Netzknoten zugeordnet und über ein gemeinsames Kommunikationsnetz miteinander verbunden, wie in Figur 8.1 dargestellt. Um sensitive Daten der jeweiligen Nutzer zu schützen und trotzdem das Netz effizient zu regeln, erstellen sie Leistungsdiagramme aus den lokal verfügbaren flexiblen Leistungen, welche sie dann untereinander austauschen. Dies ermöglicht eine Berechnung der Netzinteraktion, ohne dass die Regler Details zum aktuellen Verbrauch der Photovoltaik Einspeisung oder möglicher installierter Batteriespeicher kennen müssen. Die für die Regelung notwendigen Informationen werden sozusagen "verschlüsselt", wodurch die Privatsfähre der Nutzer geschützt wird.
Das Regelungsverfahren nutzt die Kapazitäten des Niederspannungsnetzes maximal aus und verhindert die Überlastungen der Betriebsmittel.

Das Kapitel schließt mit der Evaluierung des Verfahrens anhand einer Simulation und einer praktischen Implementierung. Die Ergebnisse der Simulation zeigen, dass das Regelungsverfahren in der Lage ist, die Übertragungskapazitäten des Netzes optimal auszunutzen und gleichzeitig die Beschränkungen in Spannung und Leiterstrom einzuhalten. Die Einspeisung von regenerativ erzeugter Leistung wird hierbei favorisiert. In der praktischen Verifikation wird indes die Echtzeitfähigkeit des Regelungsalgorithmus gezeigt.

Da Niederspannungsnetze eine inhärente Asymmetrie durch einphasige Verbraucher aufweisen, stellt die Bereitstellung von Wirk- und Blindleistung zur Balancierung der Spannungen eine wichtige Systemdienstleistung dar. In **Kapitel 5** wurde zur Behandlung dieser Problematik ein weiteres Verfahren entwickelt, welches die lokalen Umrichter nutzt, um zu einem gewissen Grad asymmetrische Leistungen einzuspeisen. Der approximierte Lastfluss aus **Kapitel 2** wird in einer dreiphasigen Formulierung genutzt, um

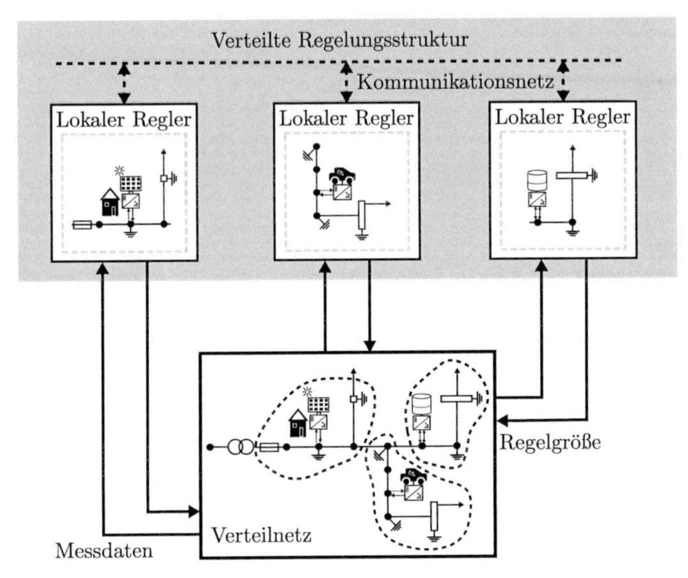

Abbildung 8.1: Grundätzliche Struktur des Netzes mit lokalen Reglern, Kommunikationsnetzwerk (rot gepunktete Linie) und direkte Messdatenaufnahme (blaue Linie) der verteilten Regelung. Es wird angenommen, dass zusätzlich intelligente Stromzähler an jedem Netzknoten angebracht sind, der keinen Regler besitzt

einen linearen Zusammenhang zwischen eingespeister Leistung und den symmetrischen Komponenten der Spannung herzustellen. Die Besonderheit dieses Ansatzes ist, dass die Komponenten des Null- und Gegensystems einzeln oder in Kombination reduziert werden können. Dies ermöglicht eine optimale Ausnutzung der Systemkomponenten, welche durch die einzelnen Regler zur Balancierung genutzt werden können.
Die grundsätzliche Idee des Ansatzes ist, dass jeder Regler zunächst die Asymmetrie minimiert, die er auch in das Netz einbringt. Dies würde jedoch bei wenigen geregelten Knoten nicht den gewünschten Effekt erzielen. Deshalb werden mittels eines verteilten Jacobi Algorithmus die lokalen Regler so koordiniert, dass sie gemeinsam die Asymmetrie im Netz reduzieren. Über das vorgestellte Kommunikationsnetzwerk werden zusätzliche Messdaten von ungeregelten Knoten ausgetauscht und in den lokalen Optimierungsproblemen berücksichtigt.

Oft sind in Niederspannungsnetzen zwar die verlegten Kabeltypen und die Struktur des Netzes, aber nicht deren genaue Impedanzwerte bekannt. Des Weiteren ändert sich der

Wirkwiderstand der Niederspannungskabel infolge der Erwärmung durch den Betriebsstrom. Diese Änderung kann im Extremfall bis zu 20 % des nominellen Wertes aus dem Datenblatt betragen. Niederspannungskabel vom Typ N(A)YY, welche in dieser Arbeit untersucht wurden, haben eine zulässige dauerhafte Betriebstemperatur von 70 °C. Diese beiden Faktoren führen dazu, dass für das approximierte Lastflussverfahren sowohl eine Schätzung der mittleren Parameter der Niederspannungskabel, als auch ein temperaturabhängiges Laufzeitmodel benötigt wird. In **Kapitel 6** werden hierfür ein Schätzalgorithmus basierend auf der Methode der kleinsten Quadrate und ein dynamisches thermisches Model der Kabel entwickelt. Es wird zudem darauf geachtet, dass die Performanz des Algorithmus auch in besonders asymmetrischen Betriebszuständen zufriedenstellend ist.

A Conductor heating evaluation

The cables are acted on with their rated current and two thirds of that. The material and geometric parameters used for the simulation are summarized in Table A.1. The soil temperature in the buried depth of $\approx 60\,\mathrm{cm}$ is set to $20\,°\mathrm{C}$ which is a good approximation for the summer period in Germany [PIK17].

Table A.1: Geometric [Nex] and material parameters used for the thermal simulation of the LV underground cables

Material	Thermal conductivity $^{W}/_{mK}$	Density $^{kg}/_{m^3}$	Heat capacity $^{J}/_{kgK}$
PVC	0.19	1300	900
Aluminum	204	2707	896
Soil	0.833-1.43	1040	∞
Cable type	Outer radius [mm]	Insulation thickness [mm]	Sheath thickness [mm]
$4 \times 240\,\mathrm{mm}^2$	26-28	2.0	2.7
$4 \times 150\,\mathrm{mm}^2$	21.5-24	1.8	2.5
$4 \times 120\,\mathrm{mm}^2$	20-22.5	1.6	2.4
$4 \times 95\,\mathrm{mm}^2$	18-20.5	1.6	2.2
$4 \times 70\,\mathrm{mm}^2$	16-18.5	1.4	2.1

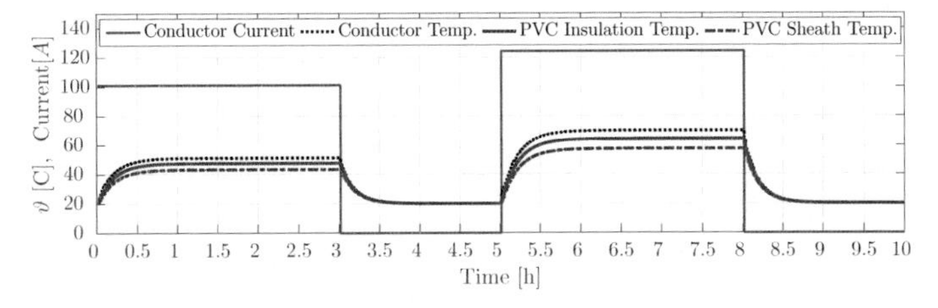

Figure A.1: Evaluation of the thermal response of a buried NAYY $4\times95\,\mathrm{mm}^2$ low voltage cable to a load current of 215 A. Between $t = 0\,\mathrm{h} - 3\,\mathrm{h}$ phase a,b and $t = 5\,\mathrm{h} - 8\,\mathrm{h}$ phase abc

B Detailed parameter data for the benchmark grids

B.1 Electrical parameters of the benchmark grid and demonstrators

Table B.1: Substation and branch parameters for the benchmark grids

Component	Type	Impedance	Max. current
Transformer A	20/0.4 kV Dyn11 630 kVA	$R_\mathrm{k} = 7.86\,\mathrm{m\Omega}$ $X_\mathrm{k} = 6.42\,\mathrm{m\Omega}$	910 A
Transformer B	20/0.4 kV Dyn11 400 kVA	$R_\mathrm{k} = 9.86\,\mathrm{m\Omega}$ $X_\mathrm{k} = 6.42\,\mathrm{m\Omega}$	577 A
Cable A	NAYY 4x6 mm^2 0.6/1 kV	$R_\mathrm{B}' = \frac{3080\,\mathrm{m\Omega}}{km}$ $X_\mathrm{B}' = \frac{90.04\,\mathrm{m\Omega}}{km}$	50 A
Cable B	NAYY 4x16 mm^2 0.6/1 kV	$R_\mathrm{B}' = \frac{1150\,\mathrm{m\Omega}}{km}$ $X_\mathrm{B}' = \frac{80.1\,\mathrm{m\Omega}}{km}$	99 A
Cable C	NAYY 4x35 mm^2 0.6/1 kV	$R_\mathrm{B}' = \frac{727\,\mathrm{m\Omega}}{km}$ $X_\mathrm{B}' = \frac{80.1\,\mathrm{m\Omega}}{km}$	121 A
Cable D	NAYY 4x50 mm^2 0.6/1 kV	$R_\mathrm{B}' = \frac{641\,\mathrm{m\Omega}}{km}$ $X_\mathrm{B}' = \frac{80.7\,\mathrm{m\Omega}}{km}$	150 A
Cable E	NAYY 4x70 mm^2 0.6/1 kV	$R_\mathrm{B}' = \frac{443\,\mathrm{m\Omega}}{km}$ $X_\mathrm{B}' = \frac{82.2\,\mathrm{m\Omega}}{km}$	185 A
Cable F	NAYY 4x95 mm^2 0.6/1 kV	$R_\mathrm{B}' = \frac{321\,\mathrm{m\Omega}}{km}$ $X_\mathrm{B}' = \frac{82.9\,\mathrm{m\Omega}}{km}$	220 A
Cable G	NAYY 4x120 mm^2 0.6/1 kV	$R_\mathrm{B}' = \frac{321\,\mathrm{m\Omega}}{km}$ $X_\mathrm{B}' = \frac{82.9\,\mathrm{m\Omega}}{km}$	220 A
Cable H	NAYY 4x150 mm^2 0.6/1 kV	$R_\mathrm{B}' = \frac{321\,\mathrm{m\Omega}}{km}$ $X_\mathrm{B}' = \frac{82.9\,\mathrm{m\Omega}}{km}$	220 A
Cable I	NAYY 4x240 mm^2 0.6/1 kV	$R_\mathrm{B}' = \frac{321\,\mathrm{m\Omega}}{km}$ $X_\mathrm{B}' = \frac{82.9\,\mathrm{m\Omega}}{km}$	220 A

Table B.2: Branch cable parameters of the 4 node LV grid

Branch number	**Cable type**	**Start - end**	**Length**
1	I	2-3	150m
2	H	3-4	250m
3	H	4-5	750m
4	H	5-6	50m

Table B.3: Branch cable parameters of the CIGRÉ residential LV grid

Branch number	**Cable type**	**Start - end**	**Length**
1	G	R1-R2	35m
2	G	R2-R3	35m
3	A	R3-R11	30m
4	D	R3-R4	35m
5	E	R4-R14	3*35m
6	D	R14-R15	30m
7	E	R4-R6	2*35m
8	B	R6-R16	30m
9	E	R6-R9	3*35m
10	A	R9-R17	30m
11	E	R9-R10	35m
12	B	R10-R18	30m

Table B.4: House load curve parameters of the CIGRÉ residential LV grid from the IEEE European LV feeder

Node number	Connected phase	Load profile	$\cos(\varphi)$	Scaling factor
1	a, b, c	3, 4, 5	0.95	1, 1, 1
2	a, b, c	6, 7, 8	0.95	1, 1, 1
3	a, b, c	9, 10, 11	0.95	1, 1, 1
4	a, b, c	12, 13, 14	0.95	1, 1, 1
5	a, b, c	15, 16, 17	0.95	1, 1, 1
6	a, b, c	18, 19, 20	0.95	1, 1, 1
7	a, b, c	21, 22, 23	0.95	1, 1, 1
8	a, b, c	24, 25, 26	0.95	1, 1, 1
9	a, b, c	27, 28, 29	0.95	1, 1, 1
10	a, b, c	30, 31, 32	0.95	1, 1, 1
11	a, b, c	33, 34, 35	0.95	1, 1, 1
12	a, b, c	36, 37, 38	0.95	1, 1, 1

Table B.5: Branch cable parameters of the 40 node LV grid

Branch number	Cable type	Start - end	Length
1	I	2-3	5m
2	H	3-4	32m
3	H	4-5	37m
4	H	5-6	41m
5	H	6-7	36m
6	G	7-8	550m
7	F	7-9	45m
8	F	9-10	42m
9	F	10-11	43m
10	F	11-12	38m
11	F	12-13	39m
12	F	13-14	34m
13	F	14-15	45m
14	I	2-16	5m
16	G	16-17	39m
17	G	17-18	38m
18	G	18-19	36m
19	G	16-20	34m
20	G	21-21	34m
21	G	21-22	39m
22	G	22-23	45m
23	G	23-24	40m
24	G	16-25	43m
25	G	25-26	44m
26	G	26-27	45m
27	G	27-28	46m
28	G	28-29	40m
29	I	2-30	5m
30	F	30-31	39m
31	F	31-32	42m
32	F	32-33	37m
33	F	33-34	38m
34	F	31-35	44m
35	F	35-36	39m
36	F	36-37	46m
37	F	37-38	41m
38	F	38-39	42m
39	F	39-40	32m

Table B.6: House load curve parameters of the 40 nodes LV grid from the IEEE European LV feeder

Node number	**Connected phase**	**Load profile**	**cos(φ)**	**Scaling factor**
1	a,b,c,n	1,51,16	0.95	1.25,1.0,0.75
2	a,b,c,n	2,52,17	0.95	1.25,1.0,0.75
3	a,b,c,n	3,53,18	0.95	1.25,1.0,0.75
4	a,b,c,n	4,54,19	0.95	1.25,1.0,0.75
5	a,b,c,n	5,55,20	0.95	1.25,1.0,0.75
6	a,b,c,n	6,56,21	0.95	1.25,1.0,0.75
7	a,b,c,n	7,57,22	0.95	1.25,1.0,0.75
8	a,b,c,n	8,58,23	0.95	1.25,1.0,0.75
9	a,b,c,n	9,59,24	0.95	1.25,1.0,0.75
10	a,b,c,n	10,60,25	0.95	1.25,1.0,0.75
11	a,b,c,n	11,61,26	0.95	1.25,1.0,0.75
12	a,b,c,n	12,62,27	0.95	1.25,1.0,0.75
13	a,b,c,n	13,63,28	0.95	1.25,1.0,0.75
14	a,b,c,n	14,64,29	0.95	1.25,1.0,0.75
15	a,b,c,n	15,65,30	0.95	1.25,1.0,0.75
16	a,b,c,n	16,66,31	0.95	1.25,1.0,0.75
17	a,b,c,n	17,67,32	0.95	1.25,1.0,0.75
18	a,b,c,n	18,68,33	0.95	1.25,1.0,0.75
19	a,b,c,n	19,69,34	0.95	1.25,1.0,0.75
20	a,b,c,n	20,70,35	0.95	1.25,1.0,0.75
21	a,b,c,n	21,71,36	0.95	1.25,1.0,0.75
22	a,b,c,n	22,72,37	0.95	1.25,1.0,0.75
23	a,b,c,n	23,73,38	0.95	1.25,1.0,0.75
24	a,b,c,n	24,74,39	0.95	1.25,1.0,0.75
25	a,b,c,n	25,75,40	0.95	1.25,1.0,0.75
26	a,b,c,n	26,75,41	0.95	1.25,1.0,0.75
27	a,b,c,n	27,76,42	0.95	1.25,1.0,0.75
28	a,b,c,n	28,77,43	0.95	1.25,1.0,0.75
29	a,b,c,n	29,78,44	0.95	1.25,1.0,0.75
30	a,b,c,n	30,79,45	0.95	1.25,1.0,0.75
31	a,b,c,n	31,80,46	0.95	1.25,1.0,0.75
32	a,b,c,n	32,81,47	0.95	1.25,1.0,0.75
33	a,b,c,n	33,82,48	0.95	1.25,1.0,0.75
34	a,b,c,n	34,83,49	0.95	1.25,1.0,0.75
35	a,b,c,n	35,84,50	0.95	1.25,1.0,0.75
36	a,b,c,n	36,85,51	0.95	1.25,1.0,0.75
37	a,b,c,n	37,86,52	0.95	1.25,1.0,0.75
38	a,b,c,n	38,87,53	0.95	1.25,1.0,0.75
39	a,b,c,n	39,89,54	0.95	1.25,1.0,0.75

B.1.1 Unbalanced setup for the validation: 40 node LV distribution grid

Table B.7: Single-phase electric vehicle charging at the 40 nodes LV grid

Node number	Connected phase	Rated power
1	–	–
2	–	–
3	–	–
4	–	–
5	a,n	2*3.6 kW
6	a,n	2*3.6 kW
7	–	–
8	–	–
9	–	–
10	–	–
11	–	–
12	a,n	4*3.6 kW
13	–	–
14	a,n	3.6 kW
15	–	–
16	–	–
17	c,n	3.6 kW
18	c,n	4*3.6 kW
19	–	–
20	c,n	2*3.6 kW
21	–	–
22	c,n	2*3.6 kW
23	–	–
24	c,n	2*3.6 kW
25	–	–
26	–	–
27	c,n	2*3.6 kW
28	–	–
29	–	–
30	–	–
31	b,n	2*3.6 kW
32	–	–
33	b,n	2*3.6 kW
34	–	–
35	b,n	4*3.6 kW
36	b,n	2*3.6 kW
37	–	–
38	–	–
39	–	–

Table B.8: Single-phase photo-voltaic units at the 40 nodes LV grid

Node number	Connected phase	Rated power
1	–	–
2	–	–
3	–	–
4	–	–
5	b,n	8 kWp
6	–	–
7	–	–
8	–	–
9	b,n	6 kWp
10	b,n	11 kWp
11	–	–
12	b,n	10 kWp
13	–	–
14	–	–
15	–	–
16	–	–
17	a,n	5 kWp
18	–	–
19	a,n	6 kWp
20	–	–
21	–	–
22	–	–
23	a,n	11 kWp
24	–	–
25	–	–
26	–	–
27	a,n	12 kW
28	–	–
29	–	–
30	c,n	5 kW
31	c,n	6 kW
32	–	–
33	c,n	11 kW
34	c,n	8 kW
35	–	–
36	–	–
37	c,n	10 kW
38	–	–
39	–	–

B.1.2 Medium voltage grid line parameters of the hardware demonstrator

Table B.9: Medium voltage cable parameters

Component	Model value	Real value
Line resistance 4 km Cable Π-ESC	$R'_{\mathrm{BM}} = \frac{273\,\mathrm{m\Omega}}{km}$	$R'_{\mathrm{BR}} = \frac{300\,\mathrm{m\Omega}}{km}$
Line inductance 4 km Cable Π-ESC	$X'_{\mathrm{L,BM}} = \frac{341.25\,\mathrm{m\Omega}}{km}$	$X'_{\mathrm{L,BR}} = \frac{375\,\mathrm{m\Omega}}{km}$
Line capacitance 4 km Cable Π-ESC	$X'_{\mathrm{C,BM}} = \frac{51.72\,\mathrm{k\Omega}}{km}$	$X'_{\mathrm{C,BR}} = \frac{56.84\,\mathrm{k\Omega}}{km}$
Line resistance 8 km Cable Π-ESC	$R'_{\mathrm{BM}} = \frac{273\,\mathrm{m\Omega}}{km}$	$R'_{\mathrm{BR}} = \frac{300\,\mathrm{m\Omega}}{km}$
Line inductance 8 km Cable Π-ESC	$X'_{\mathrm{L,BM}} = \frac{341.25\,\mathrm{m\Omega}}{km}$	$X'_{\mathrm{L,BR}} = \frac{375\,\mathrm{m\Omega}}{km}$
Line capacitance 8 km Cable Π-ESC	$X'_{\mathrm{C,BM}} = \frac{51.72\,\mathrm{k\Omega}}{km}$	$X'_{\mathrm{C,BR}} = \frac{56.84\,\mathrm{k\Omega}}{km}$

Table B.10: Medium voltage impedance load parameters

Component	Model value	Real value
Load resistance	$R'_{\mathrm{LM}} = \frac{273\,\mathrm{m\Omega}}{km}$	$R'_{\mathrm{LR}} = \frac{300\,\mathrm{m\Omega}}{km}$
Load reactance	$X'_{\mathrm{LM}} = \frac{341.25\,\mathrm{m\Omega}}{km}$	$X'_{\mathrm{LR}} = \frac{375\,\mathrm{m\Omega}}{km}$

Bibliography

[AANR05] Abel-Akher, Khalid Mohamed Nor, and Abdul Halim Abdul Rashid. Improved three-phase power-flow methods using sequence components. *IEEE Transactions on Power Systems*, 20(3):1389–1397, 2005.

[ABK+15] Filip Andrén, Benoit Bletterie, Serdar Kadam, Panos Kotsampopoulos, and Christof Bucher. On the stability of local voltage control in distribution networks with a high penetration of inverter-based generation. *IEEE Transactions on Industrial Electronics*, 62(4):2519–2529, 2015.

[ADF+10] A. Augugliaro, L. Dusonchet, S. Favuzza, M. G. Ippolito, and E. Riva Sanseverino. A backward sweep method for power flow solution in distribution networks. *International Journal of Electrical Power & Energy Systems*, 32(4):271–280, 2010.

[AGM16] F. Aeiad, W. Gao, and J. Momoh. Bad data detection for smart grid state estimation. In *Proceeding of the North American Power Symposium (NAPS)*, pages 1–6, 2016.

[AL13] M. O. Ahmed and L. Lampe. Power line communications for low-voltage power grid tomography. *IEEE Transactions on Communications*, 61(12):5163–5175, 2013.

[Alz17] Efrain Bernal Alzate. *Central reactive power control for smart low-voltage distribution grids*. PhD thesis, Universität Ulm, Institut für Energiewandlung und -speicherung, 2017.

[AMK12] Y. Aihara, R. Miyazawa, and H. Koizumi. A study on the effect of the Scott transformer on the three-phase unbalance in distribution network with single-phase generators. In *3rd IEEE International Symposium on Power Electronics for Distributed Generation Systems (PEDG)*, pages 283–290, 2012.

[AMS13] M. J. E. Alam, K. M. Muttaqi, and D. Sutanto. A three-phase power flow approach for integrated 3-wire MV and 4-wire multigrounded LV networks with rooftop solar PV. *IEEE Transactions on Power Systems*, 28(2):1728–1737, 2013.

[ANPS+16] D. B. Arnold, M. Negrete-Pincetic, M. D. Sankur, D. M. Auslander, and D. S. Callaway. Model-free optimal control of VAR resources in distribution systems: an extremum seeking approach. *IEEE Transaction on Power Systems*, 31(5):3593–3593, 2016.

[AP13] Changsun Ahn and Huei Peng. Decentralized voltage control to minimize distribution power loss of microgrids. *IEEE Transactions on Smart Grid*, 4(3):1297–1304, 2013.

[APCP13] L. R. Araujo, D. R. R. Penido, S. Carneiro, and J. L. R. Pereira. A three-phase optimal power-flow algorithm to mitigate voltage unbalance. *IEEE Transactions on Power Delivery*, 28(4):2394–2402, 2013.

[ASDB16] Daniel B. Arnold, Michael Sankur, Roel Dobb, and Kyle Brady. Optimal dispatch of reactive power for voltage regulation and balancing in unbalanced distribution systems. In *Proceedings of the IEEE Power and Energy Society General Meeting*, pages 1–5, 2016.

[AWA17] Hirofumi Akagi, Edson Hirokazu Watanabe, and Maurício Aredes. *Instantaneous power theory and applications to power conditioning*. Wiley and Sons, New York, second edition, 2017.

[AYB15] A. Anurag, Y. Yang, and F. Blaabjerg. Thermal performance and reliability analysis of single-phase PV inverters with reactive power injection outside feed-in operating hours. *in IEEE Journal of Emerging and Selected Topics in Power Electronics*, 3(4):870–880, Dec. 2015.

[Ayi17] Malinwo Estone Ayikpa. Unbalanced distribution optimal power flow to minimize losses with distributed photovoltaic plants. *International Journal of Energy and Power Engineering*, 11(2):207–212, 2017.

[AYVM+09] J. Au-Yeunga, G. M. A. Vanalme, J. M. A. Myrzik, P. Karaliolios, M. Bongaerts, J. Bozelie, and W. L. Kling. Development of a voltage and frequency control strategy for an autonomous LV network with distributed generators. In *IEEE 44th International Universities Power Engineering Conference (UPEC)*, pages 1–5, 2009.

[AZG13] E. Dall Anese, H. Zhu, and G. Giannakis. Distributed optimal power flow for smart microgrids. *IEEE Transactions on Smart Grid*, 4(3):1464–1475, 2013.

[BA07] M. Bockarjova and G. Andersson. Transmission line conductor temperature impact on state estimation accuracy. In *Proceedings of the IEEE PowerTech*, pages 701–706, 2007.

[BAJ13] Peter Birkner, Julia Antoni, and Ingo Jeromin. Integrated and autonomous smart grid concept in practice – the Frankfurt experience with a focus on acitve voltage and load flow managment. In *Proceedings of the PowerTech*, pages 1–6, 2013.

[BBL17] Markus Bell, Felix Berkel, and Steven Liu. Optimal distributed balancing control for three-phase four-wire low voltage grids. In *Proceedings of the 8th IEEE Conference on Smart Grid Communications (SmartGridComm)*, pages 229–234, 2017.

[BBL18] Markus Bell, Felix Berkel, and Steven Liu. Real-time distributed control of low voltage grids with dynamic optimal power dispatch of renewable energy sources. *IEEE Transactions on Sustainable Energy*, 10(1):417–425, 2018.

[BBM+13] Saverio Bolognani, Nicoletta Bof, Davide Michelotti, Riccardo Muraro, and Luca Schenato. Identification of power distribution network topology via voltage correlation analysis. In *Proceedings of the 52nd IEEE Conference on Decision and Control*, pages 1659–1664, 2013.

[BCCC11] Saverio Bolognani, Guido Cavraro, Federico Cerruti, and Alessandro Costabeber. A linear dynamic model for microgrid voltages in presence of distributed generation. In *Proceedings of the First IEEE International Workshop on Smart Grid Modeling and Simulation (SGMS)*, pages 31–36, 2011.

[BCCZ15] S. Bolognani, R. Carli, G. Cavraro, and S. Zampieri. Distributed reactive power feedback control for voltage regulation and loss minimization. *IEEE Transactions on Automatic Control*, 60(4):966–981, 2015.

[BCJ+13] H.G. Bock, Th Carraro, W. Jäger, S. Körkel, R. Rannacher, and J. Schlöder. *Model based parameter estimation.* Springer Verlag, 2013.

[BD15] S. Bolognani and F. Dörfler. Fast power system analysis via implicit linearization of the power flow manifold. In *Proceedings of the 53rd Annual Allerton Conf. Communication, Control and Computation*, pages 402–409, 2015.

[BDE11] BDEW. Struktur des Energieverbrauchs der privaten Haushalte nach eingesetzen Energieträgern (Stand 2011). http://www.zukunfts-energie.info, 2011. Last accessed 16.12.2017.

[Ben15] Clementine Benoit. *Models for investigation of flexibility benefits in unbalanced low voltage smart grids.* PhD thesis, G2Elab - University of Grenoble Alpes, 2015.

[Ber13] Felix Berkel. Verteilte modelprädiktive Frequenzregelung in Smart-Grids. Diploma thesis, Control System Group, University of Kaiserslautern, 2013.

[BFB+16] Markus Bell, Simon Fuchs, Felix Berkel, Steven Liu, and Daniel Görges. A privacy preserving negotiation-based control scheme for low voltage grids. In *Proceedings of the IEEE International Symposium on Industrial Electronics*, pages 678–683, 2016.

[BHK+15] Cristina García Bajo, Seyedmostafa Hashemi, Soren Baekhoej Kjaer, Guangya Yang, and Jacob Oestergaard. Voltage unbalance mitigation in LV networks using three-phase PV systems. In *Proceedings of the IEEE International Conference on Industrial Technology (ICIT)*, pages 2875–2879, 2015.

[BLR+11] Sergio Bruno, Silvia Lamonaca, Giuseppe Rotondo, Ugo Stecchi, and Massimo La Scala. Unbalanced three-phase optimal power flow for smart grids. *IEEE Transactions on Industrial Electronics*, 58(10):4504–4514, 2011.

[BMTM18] Benjamin Bayer, Patrick Matschoss, Heiko Thomas, and Adela Marian. The german experience with integrating photovoltaic systems into the low-voltage grids. *Renewable Energy*, 119:129–141, 2018.

[BNA18] BNA. EEG-Registerdaten und EEG-Fördersätze. https://www.bundesnetzagentur.de, 2018. Last accessed 03.04.2018.

[BPC+11] Stephen Boyd, Neal Parikh, Eric Chu, Borja Peleato, and Jonathan Eckstein. Distributed optimization and statistical learning via the alternating direction method of multipliers. *Foundation and Trend in Machine Learning*, page 1–122, 2011.

[Bra06] K. D. Brabandere. *Voltage and frequency droop control in low volt- age grids by distributed generators with inverter front-end.* PhD thesis, Katholieke Universiteit Leuven, Faculteit Ingenieurswetenschappen, De- partement Elektrotechniek Afdeling Elektrische Energie en Computerarchi- tecturen, 2006.

[BSI19] BSI. BSI TR-03109 Technische Vorgaben für intelligente Messsysteme und deren sicherer Betrieb. https://www.bsi.bund.de, 2019. Last accessed 07.07.2019.

[BV04] Stephen Boyd and Lieven Vandenberghe. *Convex optimization.* Cambridge University Press, 2004.

[BW116] NETZlabor Niederstetten. https://www.netze-bw.de, 2016. Last accessed 03.04.2018.

[BW89] Mesut E. Baran and Felix F. Wu. Optimal sizing of capacitors placed on a radial distribution system. *IEEE Transactions on Power Delivery*, 4(1):735–743, 1989.

[BWA+17] Andrey Bernstein, Cong Wang, Emiliano Dall Anese, Jean-Yves Le Boudec, , and Changhong Zhao. Load-flow in multiphase distribution networks: existence, uniqueness, and linear models. arXiv:1702:03310v1, 2017.

[BZ11] Saverio Bolognani and Sandro Zampieri. Distributed control for optimal reactive power compensation in smart microgrids. In *Proceedings of the 50th IEEE Conference on Decision and Control and European Control Conference (CDC-ECC)*, pages 6630–6635, 2011.

[BZ15] Saverio Bolognani and Sandro Zampieri. On the existence and linear approximation of the power flow solution in power distribution networks. *IEEE Transactions on Power Systems*, 31(1):163–172, 2015.

[CA98] A. J. Conejo and J. A. Aguado. Multi-area coordinated decentralized DC optimal power flow. *Transactions on Power Systems*, 13(4):1272–1278, 1998.

[CBA07] Eduardo F. Camacho and Carlos Bordons Alba. *Model predictive control.* Springer Verlag, 2007.

[CBCZ16] G. Cavraro, S. Bolognani, R. Carli, and S. Zampieri. The value of communication in the voltage regulation problem. In *55th iEEE Confernce on Decission and Control (CDC)*, pages 5781–5786, 2016.

[CBCZ17] Guido Cavraro, Saverio Bolognani, Ruggero Carli, and Sandro Zampieri. The value of communication in the voltage regulation problem. http://people.ee.ethz.ch, 2017. Last accessed 10.02.2019.

[CC00] T.H. Chen and J.T. Cherng. Optimal phase arrangement of distribution transformers connected to a primary feeder for system unbalance improvement and loss reduction using a genetic algorithm. *Transactions on Power Systems*, 15(3):994–1000, 2000.

[CCM+18] Antonio Camacho, Miguel Castilla, Jaume Miret, García de Vicuña, and Ramon Guzman. Positive and negative sequence control strategies to maximize the voltage support in resistive-inductive grids during grid faults. *IEEE Transactions on Power Electronics*, 33(6):5362–5373, 2018.

[CFO03] Rade M. Ciric, Antonio Padilha Feltrin, and Luis F. Ochoa. Power flow in four-wire distribution networks general approach. *IEEE Transactions on Power Systems*, 18(4):1283–1290, 2003.

[CGW05] Florin Capitanescu, Mevludin Glavic, and Louis Wehenkel. An interior-point method based optimal power flow. In *Proceedings of the 3-rd ACOMEN conference*, pages 1–18, 2005.

[Cig14] Benchmark systems for network integration of renewable and distributed energy resources. Technical report, CIGRE, 2014.

[CJKT02] E. Camponogara, D. Jia, B. Krogh, and S. Talukdar. Distributed model predictive control. *IEEE Control Systems Magazine*, 22(1):44–52, 2002.

[CLT+12] K. H. Chua, Yun Seng Lim, Phil Taylor, Stella Morris, and Jianhui Wong. Energy storage system for mitigating voltage unbalance on low-voltage networks with photovoltaic systems. *IEEE Transactions on Power Delivery*, 27(4):1783–1790, 2012.

[CMCB06] E. Castillo, R. Minguez, A. Conejo, and R. Garcia Bertrand. *Decomposition techniques in mathematical programming*. Springer Verlag, Berlin, 2006.

[COC12] Mary B. Cain, Richard P. O'Neil, and Anya Castillo. History of optimal power flow and formulations. http://www.ferc.gov, 2012. Last accessed 10.03.2018.

[CS95] Carol S. Cheng and Dariush Shrimohammadi. A three-phase power flow method for real-time distribution system analysis. *Transactions on Power Systems*, 10(2):671–679, 1995.

[CSdlPL13] Panagiotis Christofides, Riccardo Scattolini, David Munoz de la Pena, and Jinfeng Liu. Distributed model predictive control: A tutorial review and future research directions. *Computers & Chemical Engineering*, 51:21–41, 2013.

[CT08] Claudio Canuto and Anita Tabacco. *Mathematical analysis I*. Springer-Verlag Milan, 2008.

[CTV+17] Mirsad Cosovic, Achilleas Tsitsimelis, Dejan Vukobratovic, Javier Matamoros, and Carles Anton-Haro. 5G mobile cellular networks: enabling distributed state estimation for smart grids. *IEEE Communications Magazine*, 55(10):62–69, 2017.

[CZMJ13] Christian Conte, Melanie N. Zeilinger, Manfred Morari, and Colin N. Jones. Robust distributed model predictive control of linear systems. In *Proceedings of the European Control Conference (ECC)*, pages 2764–2769, 2013.

[Das17] Jamini. C. Das. *Understanding symmetrical components for power system modeling*. Wiley-IEEE Press, 2017.

[DBC16] Deepjyoti Deka, Scott Backhaus, and Michael Chertkov. Learning topology of distribution grids using only terminal node measurements. In *Proceedings of the IEEE International Conference on Smart Grid Communications (SmartGridComm)*, pages 205–211, 2016.

[DE11] DIN-EN-50160:2011-02. Voltage characteristics of electricity supplied by public distribution networks. https://www.beuth.de, 2011. German version, last accessed 07.12.2017.

[Deg15] Merkebu Zenebe Degefa. *Real-time thermal state and component loading estimation in active distribution networks.* PhD thesis, Aalto University, 2015.

[DEN17] DENA. Plattform Systemdienstleistungen. http://www.plattform-systemdienstleistungen.de/, 2017. Last accessed 20.06.2018.

[DEN18] DENA. Kenndaten des deutschen Stromnetzes. https://www.dena.de, 2018. Last accessed 20.06.2018.

[DGC15] Sairaj V. Dhople, Swaroop S. Guggilam, and Yu Christine Chen. Linear approximations to AC power flow in rectangular coordinates. In *Proceedings of the 53rd Annual Allerton Conference Communication, Control and Computation*, pages 211–217, 2015.

[DGW12] E. Dall'Anese, G. B. Giannakis, and B. F. Wollenber. Optimization of unbalanced power distribution networks via semidefinite relaxation. In *IEEE North American Power Symposium (NAPS)*, pages 1–6, 2012.

[Dom13] Alexander Domahidi. *Methods and tools for embedded optimization and control.* PhD thesis, ETH Zürich, 2013.

[DPDB03] J.J. Desmet, D.J. Putman, F. D'hulster, and R.J. Belmans. Thermal analysis of the influence of nonlinear, unbalanced and asymmetric loads on current conducting capacity of LV-cables. In *Proceedings of the IEEE PowerTech Conference*, pages 8–14, 2003.

[Dug10] R. C. Dugan. Open distribution simulations system workshop: using OpenDSS for smart distribution simulations. In *Proceedings of the EPRI PQ Smart Distribution Conference and Exhibition*, 2010.

[DWS15] W. H. Wellssow J. Jordan R. Bischler D. Waeresch, R. Brandalik and N. Schneider. Linear state estimation in low voltage grids based on smart meter data. In *IEEE Eindhoven PowerTech*, pages 1–6, 2015.

[ER99] Antonio Gomez Exposito and Esther Romero Ramos. Reliable load flow technique for radial distribution networks. *IEEE Transactions on Power Systems*, 14(3):1063–1069, 1999.

[ERG11] F. M. Echavarren, L. Rouco, and A. Gonzalez. Dynamic thermal mode model of isolated cables. In *Proceedings of the 17th Power Systems Computation Conference*, 2011.

[Ers14] T. Erseghe. Distributed optimal power flow using ADMM. *IEEE Transactions on Power Systems*, 29(5):2370–2380, 2014.

[ETV13] T. Erseghe, S. Tomasin, and A. Vigato. Topology estimation for smart micro grids via powerline communications. *IEEE Transactions on Signal Processing*, 61(13):3368–3377, 2013.

[FMXY12] Xi Fang, Satyajayant Misra, Guoliang Xue, and Dejun Yang. Smart grid – the new and improved power grid: a survey. *IEEE Communications Surveys and Tutorials*, 14(4):944–980, 2012.

[Fog11] Efi Fogel. Minkowski sum—definition, complexity, construction, applications. http://acg.cs.tau.ac.il, March 2011. Last accessed 10.06.2019.

[For17] Philipp Fortenbacher. *On the integration of distributed battery storage in low voltage grids*. PhD thesis, ETH Zürich, 2017.

[FR16] Stephen Frank and Steffen Rebennack. An introduction to optimal power flow: theory, formulation, and examples. *IIE Transactions*, 48(12):1172–1197, 2016.

[Fra17] Fraunhofer. Installed renewable generation capacity. https://www.energy-charts.de/, July 2017. Last accessed 10.05.2018.

[FSJ14] F. Farokhi, I. Shames, and K.H. Johansson. *Distributed MPC via dual decomposition and alternative direction method of multipliers*. Springer Dordrecht, 2014. pp. 115-113.

[FSR12] Stephen Frank, Ingrida Steponavice, and Steffen Rabennack. *Optimal power flow: a bibliographic survey I, formulations and deterministic methods*, pages 221–258. Springer Verlag, 2012.

[FSZ10] A. Ghosh G. Ledwich F. Shahnia, R. Majumder and F. Zare. Sensitivity analysis of voltage imbalance in distribution networks with rooftop PVs. In *IEEE PES General Meeting, Minneapolis, MN*, pages 1–8, 2010.

[GAC$^+$16] Swaroop S. Guggilam, Emiliano Dall Anese, Yu Christine Chen, Sairaj V. Dhople, and Georgios B. Giannakis. Scalable optimization methods for distribution networks with high PV integration. *IEEE Transactions on Smart Grid*, 7(4):2061–2069, 2016.

[Gao13] Chao Gao. *Voltage control in distribution networks using on-load tap changer transformers*. PhD thesis, University of Bath, The Department of Electronic and Electrical Engineering, 2013.

[Gar15] Alejandro Garces. A quadratic approximation for the optimal power flow in power distribution systems. *Electric Power Systems Research*, 130:222–229,

April 2015.

[Gar16] Alejandro Garces. A linear three-phase load flow for power distribution systems. *IEEE Transactions on Power Systems*, 31(1):827–828, 2016.

[GDC+16] Swaroop S. Guggilam, Emiliano Dall'Anese, Yu Christine Chen, Sairaj V. Dhople, and Georgios B. Giannakis. Scalable optimization methods for distribution networks with high PV integration. *IEEE Transactions on Smart Grid*, 7(4):2061–2069, 2016.

[GDK+13] Pontus Giselsson, Minh Dang Doan, Tamás Keviczky, Bart De Schutter, and Anders Rantzer. Accelerated gradient methods and dual decomposition in distributed model predictive control. *Automatica*, 49:829–833, 2013.

[Gie13] P. Gieselsson. Output feedback distributed model predictive control with inherent robustness properties. In *American Control Confernce (ACC)*, pages 1691–1696, 2013.

[GKA14] Simon Gill, Ivana Kockar, and Graham W. Ault. Dynamic optimal power flow for active distribution networks. *IEEE Transactions on Power Systems*, 29(1):121–131, 2014.

[GL14] Lingwen Gan and Steven H. Low. Convex relaxations and linear approximation for optimal power flow in multiphase radial networks. In *Proceedings of the 18th Power Systems Computation Conference (PSCC)*, pages 1–9, 2014.

[GL16] Lingwen Gan and Steven H. Low. An online gradient algorithm for optimal power flow on radial networks. *IEEE Journal of Selected Areas in Communication*, 34(3):625–639, 2016.

[GMR19] Kshitij Girigoudar, Daniel K. Molzahn, and Line A. Roald. Analytical and empirical comparisons of voltage unbalance definitions. Last accessed: 26.06.2019 20:18, 2019.

[Goe16] Andreas Goetz. *Zukünftige Belastungen von Niederspannungsnetzen unter besonderer Berücksichtigung der Elektromobilität.* PhD thesis, Technische Universität Chemnitz, 2016.

[HBH16] Adrian Hauswirth, Saverio Bolognani, Gabriela Hug, and Florian Dörfler 1. Projected gradient descent on riemannian manifolds with applications to online power system optimization. In *In Proceedings 54rd Annual Allerton Conference on Communication, Control, and Computing*, pages 225–232, 2016.

[HEAF10] M. Hassanzadeh, M. Etezadi-Amoli, and M. S. Fadali. Practical approach for sub-hourly and hourly prediction of pv power output. In *In Proceedings of the North American Power Symposium*, pages 1–5, 2010.

[HJL+12] Ralph M. Hermans, Andrej Jokic, Mircea Lazar, Alessandro Alessio, Paul P.J. Van den Bosch, Ian A. Hiskens, and Alberto Bemporad. Assessment of non-centralised model predictive control techniques for electrical power networks. *International Journal of Control*, 85(8):1162–1177, 2012.

[HL17] Dale Hall and Nic Lutsey. Emerging best practices for electric vehicle charging infrastructure. https://www.theicct.org, 2017. Last accessed 10.02.2018.

[JK01] D. Jia and B.H. Krogh. Distributed model predictive control. In *American Control Conference (ACC)*, pages 2767–2772, 2001.

[JSSM14] Dongli Jia, Wanxing Sheng, Xiaohui Song, and Xiaoli Meng. A system identification method for smart distribution grid. In *Proceedings of the International Conference on Power System Technology (POWERCON)*, pages 14–19, 2014.

[Kar] Helmut Karger. Info letter No. 5: zero sequence voltage in three-phase networks. https://www.a-eberle.de. Last accessed 05.06.2017.

[Kau95] W. Kaufmann. *Planung öffentlicher Elektrizitätsversorgungssysteme*. VDE-Verlag, 1995.

[KCK13] Theodoros Kyriakidis, Rachid Cherkaoui, and Maher Kayal. A DC power flow extension. In *Proceedings of the IEEE PES Innovative Smart Grid Technologies Europe (ISGT Europe)*, pages 1–5, 2013.

[KCLB13] M. Kraning, E. Chu, J. Lavaei, and S. Boyd. Dynamic network energy management via proximal message passing. *Foundation and Trend in Machine Learning*, 1(2):73–126, 2013.

[Ken17] Fabian Kennel. *Beitrag zu iterativ lernenden modellprädiktiven Regelungen*. PhD thesis, Technische Universität Kaiserslautern, Lehrstuhl für Regelungssysteme, 2017.

[Ker08] W. H. Kersting. A three-phase unbalanced line model with grounded neutrals through a resistance. In *Proceeding of the IEEE Power and Energy Society General Meeting - Conversion and Delivery of Electrical Energy in the 21st Century*, pages 1–2, 2008.

[Ker10] Georg Kerber. *Aufnahmefähigkeit von Niederspannungsverteilnetzen für die Einspeisung aus Photovoltaikkleinanlagen*. PhD thesis, Technische Universität München, 2010.

[Ker12] W. H. Kersting. *Distribution system modeling and analysis.* CRC Press, 2012.

[KT05] Willett Kempton and Jasna Tomic. Vehicle-to-grid power implementation: From stabilizing the grid to supporting large-scale renewable energy. *Journal of Power Sources*, 144(1):280–294, 2005.

[KZGB16] Vassilis Kekatos, Liang Zhang, Georgios B. Giannakis, and Ross Baldick. Voltage regulation algorithms for multiphase power distribution grids. *IEEE Transactions on Power Systems*, 31(5):3913–3923, 2016.

[Lan13] Nils Neusel Lange. *Dezentrale Zustandsüberwachung für intelligente Niederspannungsnetze.* PhD thesis, Bergische Universität Wuppertal, 2013.

[Lar00] Mats Larsson. *Coordinated voltage control in electric power systems.* PhD thesis, Lund University, Department of Industrial Electrical Engineering and Automation, 2000.

[LC13] Yun Li and Peter A. Crossley. Voltage balancing in low voltage distribution networks using Scott transformers. In *CIRED 22nd International Conference on Electricity Distribution*, pages 1–4, June 2013.

[Leu09] Udo Leuschner. Die Entwicklung der deutschen Stromversorgung bis 1998. Technical report, 2009.

[LK09] Yuan Liao and Mladen Kezunovic. Online optimal transmission line parameter estimation for relaying applications. *IEEE Transactions on Power Delivery*, 24(1):96–102, 2009.

[LO16] Chao Long and Luis F. Ochoa. Voltage control of PV-rich LV networks: Oltc-fitted transformer and capacitor banks. *IEEE Transactions on Power Systems*, 31(5):4016–4025, 2016.

[LPS13] Xiao Li, H. Vincent Poor, and Anna Scaglione. Blind topology identification for power systems. In *Proceedings of the IEEE International Conference on Smart Grid Communications (SmartGridComm)*, pages 91–96, 2013.

[LTP07] R. Somerville U. Cubasch Y. Ding C. Mauritzen A. Mokssit T. Peterson Le Treut, H. and M. Prather. *Historical overview of climate change science*, volume Climate Change 2007: The Physical Science Basis. Contribution of Working Group I to the Fourth Assessment Report of the Intergovernmental Panel on Climate Change, chapter 1, pages 93–127. Cambridge University Press, Cambridge, United Kingdom and New York, NY, USA, 2007.

[LTS16] L. Lampe, A. M. Tonello, and T. G. Swart. *Power line communications:*

principles, standards and applications from multimedia to smart-grid. Wiley, 2 edition, 2016.

[Lun16] Jan Lunze. *Regelungstechnik 2 - Mehrgrößensysteme, Digitale Regelung.* Springer-Verlag Berlin Heidelberg, 2016.

[LWX12] Jing Li, Wei Wei, and Ji Xiang. A simple sizing algorithm for stand-alone PV/wind/battery hybrid microgrids. *Energies*, (5):5307–5323, 2012.

[MCR15] Ognjen Marjanovic Martin Caton and Simon Rowland. Dynamic thermal modelling of low voltage underground cables. Technical report, University of Manchester, 2015.

[MDS+17] Daniel K. Molzahn, Florian Dörfler, Henrik Sandberg, Steven H. Low, Sambuddha Chakrabarti, Ross Baldick, and Javad Lavaei. A survey of distributed optimization and control algorithms for electric power systems. *IEEE Transactions on Smart Grid*, 8(6):2941–2962, 2017. Accepted for Publication.

[MHR12] D. Montenegro, M. Hernandez, and G. A. Ramos. Real time OpenDSS framework for distribution systems simulation and analysis. In *Proceeding of the 6th IEEE/PES Transmission and Distribution: Latin America Conference and Exposition (T&D-LA)*, pages 1–5, 2012.

[Mic17] ST Microelectronics. 2 kW 3-phase motor control evaluation board featuring the STGIPS20C60 IGBT intelligent power module. http://www.st.com, 2017. Last accessed 09.11.2017.

[Mil06] Robert John Millar. *A comprehensive approach to real-time power cable temperature prediction and rating in thermally unstable environments.* PhD thesis, Helsinki University of Technology, 2006.

[Mil10] Frederico Milano. *Power system modelling and scripting.* Springer Verlag, 2010.

[ML99] M. Moari and H. J. Lee. Model predictive control: past, present and future. In *Coputers and Chemical Engineering*, volume 23, pages 667–682, 1999.

[MN14] Jose M. Maestre and Rudy R. Negenborn. *Distributed MPC made easy.* Springer Verlag, 2014.

[Moh13] Padideh Ghafoor Mohseni. *Coordination techniques for distributed model predictive control.* PhD thesis, University of Alberta, Department of Chemical and Materials Engineering, 2013.

[Mon18] Pablo R. Baldivieso Monasterios. *Distributed model predictive control for reconfigurable large-scale systems.* PhD thesis, Department of Automatic

Control and Systems Engineering, University of Sheffield, 2018.

[Mor13] Mohammad Moradzadeh. *Voltage coordination in multi-area power systems via distributed model predictive control.* PhD thesis, University Gent, 2013.

[Mos16] Abolfazl Mosaddegh. *Optimal operation of power distribution feeders with smart loads.* PhD thesis, University of Waterloo, 2016.

[Mou18] Gautham Ram Chandra Mouli. *Charging electric vehicles from solar energy Power converter, charging algorithm and system design.* PhD thesis, Delft University of Technology, 2018.

[MRRS00] D. Mayne., B. Rawlings, C. Rao, and P. Scokaert. Constraint model predictive control: Stability and optimality. *Automatica*, 36(6):789–814, 2000.

[MSR05] D. Q. Mayne, M.M. Seron, and S. V. Rakovic. Robust model predictive control of constrained linear systems with bounded disturbances. *Automatica*, 41(2):219–224, 2005.

[MWGC16] W. J. Ma, J. Wang, V. Gupta, and C. Chen. Distributed energy management for networked microgrids using online alternating direction method of multipliers with regret. *IEEE Transactions on Smart Grid*, 9(2):847–856, 2016.

[Neg07] Rudy R. Negenborn. *Multi-agent model predictive control with applications to power networks.* PhD thesis, Technische Universiteit Delft, Dutch Institute of Systems and Control, 2007.

[Nex] Nexans. NAYY-O low voltage cable parameters. http://www.nexans.de. Last accessed 06.06.2018.

[Nie16] Olli Niemitalo. C++ code for ellipse–circle and ellipse–ellipse collision detection. http://yehar.com, 11 2016. Last accessed 04.05.2019.

[NN15] Björn Nykvist and Mans Nilsson. Rapidly falling costs of battery packs for electric vehicles. *Nature Climate Change*, (5):329–332, 2015.

[NND11] Ion Necorara, Valentin Nedelcu, and Ioan Dumitrache. Parallel and distributed optimization methods for estimation and control in networks. *Journal of Process Control*, 21(5):756–766, 2011.

[Oer14] Christian Oerter. *Autarke, koordinierte Spannungs- und Leistungsregelung in Niederspannungsnetzen.* PhD thesis, Bergische Universität Wuppertal, 2014.

[OLS+13] Christian Oerter, Nils Neusel Lange, Philipp Sahm, Markus Zdrallek, Wolfgang Friedrich, and Martin Stiegler. Experience with first smart, autonomous LV-grids in germany. In *Proceedings of the 22nd International Conference on Electricity Distribution (CIRED)*, pages 1–4, 2013.

[OO04] Dietrich Oeding and Bernd Rüdiger Oswald. *Elektrische Kraftwerke und Netze*. Springer Verlag, 2004.

[PBPR17] Satya Jayadev Pappu, Nirav Bhatt, Ramkrishna Pasumarthy, and Aravind Rajeswaran. Identifying topology of low voltage distribution networks based on smart-meter data. *IEEE Transactions on Smart Grid*, 9(5):5113–5122, 2017.

[Pep16] Jouni Peppanen. *Improving distribution system model accuracy by leveraging ubiquitous sensors*. PhD thesis, Georgia Institute of Technology, 2016.

[PES15] PES. Distribution test feeders. https://ewh.ieee.org, 2015. Last accessed 13.04.2018.

[PGK+16] Alexander M. Prostejovsky, Oliver Gehrke, Anna M. Kosek, Thomas Strasser, and Henrik W. Bindner. Distribution line parameter estimation under consideration of measurement tolerances. *IEEE Transactions on Industrial Informatics*, 12(2):726–735, 2016.

[PGRB16a] Jouni Peppanen, Santiago Grijalva, Matthew J. Reno, and Robert J. Broderick. Distribution system low-voltage circuit topology estimation using smart-metering data. In *Proceedings of the IEEE PES Transmission and Distribution Conference and Exposition (T&D)*, pages 1–5, 2016.

[PGRB16b] Jouni Peppanen, Santiago Grijalva, Matthew J. Reno, and Robert J. Broderick. Secondary circuit model generation using limited PV measurements and parameter estimation. In *Proceedings of the Power and Energy Society General Meeting (PESGM)*, pages 1–5, 2016.

[PHS05] Stavros Papathanassiou, Nikos Hatziargyriou, and Kai Strunz. A benchmark low voltage microgrid network. In *Proceedings of the CIGRE Symposium: Power Systems with Dispersed Generation, Technologies, Impacts on Development, Operation and Performances*, pages 313–321, 2005.

[PIK17] PIK. Soil temperatur measurements. https://www.pik-potsdam.de, 2017. Last accessed 12.12.2017.

[PL15] Qiuyu Peng and Steven H. Low. Distributed algorithm for optimal power flow on an unbalanced radial network. In *Proceedings of the IEEE 54th Annual Conference on Decision and Control (CDC)*, pages 6915–6920, 2015.

[PM01] P. Pillay and M. Manyage. Definitions of voltage unbalance. *IEEE Power Engineering Review Magazine*, 5:50–51, May 2001.

[PRK+13] Sophie Pelland, Jan Remund, Jan Kleissl, Takashi Oozeki, and Karel De Brabandere. Photovoltaic and solar forecasting: state-of-the-art. Technical report, International Energy Agency, 2013.

[PT17] F. Passerini and A. M. Tonello. On the exploitation of admittance measurements for wired network topology derivation. *IEEE Transactions on Instrumentation and Measurements*, 66(3):374–382, 2017.

[Qua18] Volker Quasching. Statistiken: Installierte Photovoltaikleistung in Deutschland. https://www.volker-quaschning.de, 2018. Last accessed 03.04.2018.

[RBT17] W. H. Wellssow R. Brandalik, D. Waeresch and J. Tu. Linear three-phase state estimation for LV grids using pseudo-measurements based on approximate power distributions. *CIRED - Open Access Proceedings Journal*, 10(1):1871–1874, 2017.

[REÁC04] Esther Romero Ramos, Antonio Gómez Expósito, and Gabriel Álvarez Cordero. Quasi-coupled three-phase radial load flow. *IEEE Transactions on Power Systems*, 19(2):776–781, 2004.

[RKK19] Bharath Varsh Rao, Friederich Kupzog, and Martin Kozek. Three-phase unbalanced optimal power flow using holomorphic embedding load flow method. *Sustainability*, 11(6), 2019. 1774.

[Roa16] Line Alnaes Roald. *Optimization methods to manage uncertainty and risk in power systems operation*. PhD thesis, ETH Zürich, 2016.

[RS08] J.B. Rawlings and B.T. Stewart. Coordinating multiple optimization-based controllers: New opportunities and challenges. *Journal of Process Control*, 18(9):839–845, 2008.

[Sca09] R. Scattolini. Architectures for distributed and hierarchical model predictive control - a review. *Journal of Process Control*, 19(5):723–731, 2009.

[Sch15a] Dierk Schröder. *Elektrische Antriebe - Regelung von Antriebssystemen*. Springer Verlag, 2015.

[Sch15b] Adolf J. Schwab. *Elektroenergiesysteme: Erzeugung, Übertragung und Verteilung elektrischer Energie*. Springer Verlag, 2015.

[Sch17] Robert Schwerdfeger. *Vertikaler Netzbetrieb - Ein Ansatz zur Koordination von Netzbetriebsinstanzen verschiedener Netzebenen*. PhD thesis, Technische Universität Ilmenau, 2017.

[SGLZ11] F. Shahnia, A. Ghosh, G. Ledwich, and F. Zare. Voltage unbalance sensitivity analysis of plug–in electric vehicles in distribution networks. In *21st Australasian Universities Power Engineering Conference (AUPEC)*, pages

1–6, 2011.

[Sha14] Isha Sharma. *Operation of distribution systems with PEVs and smart-loads.* PhD thesis, University of Waterloo, 2014.

[Sho04] T. A. Short. *Electric power distribution handbook.* CRC Press, 2004.

[SHSL88] D. Shirmohammadi, H.W. Hong, A. Semlyen, and G. X. Luo. A compensation-based power flow method for weakly meshed distribution and transmission networks. *IEEE Transactions on Power Systems*, 3(2):753–762, 1988.

[Sim06] Dan Simon. *Optimal state estimation: Kalman, H-infinity, and nonlinear approaches.* Wiley, 2006.

[SJA09] B. Stott, J. Jardim, and O. Alsac. DC power flow revisited. *IEEE Transactions on Power Systems*, 24(3):1290–1300, 2009.

[SJK10] Bhim Singh, P. Jayaprakash, and D P Kothari. Magnetics for neutral current compensation in three-phase four-wire distribution system. In *Joint International Conference on Power Electronics, Drives and Energy Systems & Power*, pages 1–7, 2010.

[SMW14] Xiangjing Su, M.A.S. Masoum, and P. Wolfs. Comprehensive optimal photovoltaic inverter control strategy in unbalanced three-phase four-wire low voltage distribution networks. *IET Generation, Transmission & Distribution*, 8(11):1848–1859, 2014.

[SSM07] A.K. Singh, G.K. Singh, and R. Mitra. Some observations on definitions of voltage unbalance. In *39th North American Power Symposium (NAPS)*, pages 473–479, 2007.

[Ste13] Thomas Stetz. *Autonomous voltage control strategies in distribution grids with photovoltaic systems - technical and economic assessment.* PhD thesis, University of Kassel, Department of Energy Management and Power System Operation, 2013.

[STM17] STM. Discovery kit with STM32F407VG MCU. http://www.st.com, 2017. Last accessed 09.11.2017.

[Sto74] Brian Stott. Review of load-flow calculation methods. In *Proceedings of the IEEE*, volume 62, 1974.

[SWG14] F. Shahnia, P.J. Wolfs, and A. Ghosh. Voltage unbalance reduction in low voltage feeders by dynamic switching of residential customers among three phases. *IEEE Transactions on Smart Grid*, 5(3):1318–1327, 2014.

[Tit11] Peter Tittmann. *Graphentheorie: eine anwendungsorientierte Einführung; mit zahlreichen Beispielen und 80 Aufgaben.* Carl-Hanser-Verlag, 2011.

[Tsi13] Konstantios Tsianos. *The role of the network in distributed optimization algorithms: convergence rates, scalability, communication / computation tradeoffs and communication delays.* PhD thesis, McGill University, Department Electrical and Computer Engineering, 2013.

[UT13] Andrew J. Urquhart and Murray Thomson. Assumptions and approximations typically applied in modelling LV networks with high penetrations of low carbon technologies. In *Proceedings of the 3rd Solar Integration Workshop*, pages 1–6, 2013.

[Var06] Alex A. Kurzhanskiy Pravin Varaiya. Ellipsoidal toolbox. Technical report, Department of Electrical Engineering and Computer Sciences, University of California at Berkeley, 2006.

[VDE18] VDE. VDE-AR-N 4105 Anwendungsregel:2018-11, 11 2018.

[Ven06] Aswin Venkat. *Distributed model predictive control: theory and applications.* PhD thesis, University of Wisconsin-Madison, 2006.

[VHRW08] Aswin N. Venkat, Ian A. Hiskens, James B. Rawlings, and Stephen J. Wright. Distributed MPC strategies with application to power system automatic generation control. *IEEE Transactions on Control Systems Technology*, 16(6):1192–1206, 2008.

[VRW06] A. Venkat, J. Rawlings, and S. Wright. Stability and optimality of distributed model predictive control. In *44th IEEE Conference Decision and Control*, pages 6680–6685, 2006.

[Wir18] Harry Wirth. Recent facts about photovoltaics in germany. Technical report, Modules and Power Plants, Fraunhofer ISE, 2018.

[Woo96] W. B. Wood. *Power generation operation and control.* Wiley and Sons, New York, 1996.

[WS09] V. Wesselak and T. Schabbach. *Regenerative Energietechnik.* Springer-Verlag Berline Heidelberg, 2009.

[Ye00] Zhihong Ye. *Modeling and control of parallel three-phase PWM converters.* PhD thesis, Virginia Polytechnic Institute and State University, 2000.

[Yin11] Shaoqing Ying. *Auswirkungen auf die Niederspannungsnetze bei hoher Penetration von innerstädtischen Photovoltaikanlagen und Elektrofahrzeugen.* PhD thesis, Brandenburgischen Technischen Universität Cottbus, 2011.

[Zac17] Jan Zacharias. Konkretisierung des Ampelkonzepts im Verteilungsnetz. https://www.bdew.de, 2017. Last accessed 16.04.2019.

[ZC95] Ray D. Zimmermann and Hsiao-Dong Chiang. Fast decoupled power flow for unbalanced radial distribution systems. *IEEE Transactions on Power Systems*, 10(4):2045–2052, 1995.

[ZDGD13] B. Zhang, A. Dominguez-Garcia, and D.Tse. Local control approach to voltage regulation in distribution networks. In *Proceeding of the North American Power Symposium (NAPS)*, pages 1–6, 2013.

[Zim95] Daniel Zimmerman. *Comprehensive distribution power flow: modeling, formulation, solution, algorithms and analysis*. PhD thesis, Cornell University, 1995.

[ZLDGT14] B. Zhang, A. Lam, A. Dominguez-Garcia, and D. Ts. An optimal and distributed method for voltage regulation in power distribution system. *IEEE Transactions on Power Systems*, 30(4):1714–1726, 2014.

CURRICULUM VITAE

Personal details

Name: Markus Bell
Gender: Male
Date of birth: 21th of April, 1985
Place of birth: Tauberbischofsheim, Germany
Present citizenship: German

Education

1995–2001 Middle-school Waibstadt
2001–2004 Apprenticeship as an electrician
2004–2005 Theodor Frey school Eberbach
2006–2010 Electrical Engineering, University of Applied Sciences Mannheim (B. Sc.)
2010–2012 Electrical Engineering, Ruhr-University Bochum (M. Sc.)
2012–2018 Ph.D. candidate, University of Kaiserslautern

Work experience

2008–2009 6 month internship FGH Engineering & Test GmbH
2012–2018 Research assistant, Chair of Control Systems, University of Kaiserslautern

In der Reihe *„Forschungsberichte aus dem Lehrstuhl für Regelungssysteme"*, herausgegeben von Steven Liu, sind bisher erschienen:

1	Daniel Zirkel	Flachheitsbasierter Entwurf von Mehrgrößenregelungen am Beispiel eines Brennstoffzellensystems ISBN 978-3-8325-2549-1, 2010, 159 S. 35.00 €
2	Martin Pieschel	Frequenzselektive Aktivfilterung von Stromoberschwingungen mit einer erweiterten modellbasierten Prädiktivregelung ISBN 978-3-8325-2765-5, 2010, 160 S. 35.00 €
3	Philipp Münch	Konzeption und Entwurf integrierter Regelungen für Modulare Multilevel Umrichter ISBN 978-3-8325-2903-1, 2011, 183 S. 44.00 €
4	Jens Kroneis	Model-based trajectory tracking control of a planar parallel robot with redundancies ISBN 978-3-8325-2919-2, 2011, 279 S. 39.50 €
5	Daniel Görges	Optimal Control of Switched Systems with Application to Networked Embedded Control Systems ISBN 978-3-8325-3096-9, 2012, 201 S. 36.50 €
6	Christoph Prothmann	Ein Beitrag zur Schädigungsmodellierung von Komponenten im Nutzfahrzeug zur proaktiven Wartung ISBN 978-3-8325-3212-3, 2012, 118 S. 33.50 €
7	Guido Flohr	A contribution to model-based fault diagnosis of electro-pneumatic shift actuators in commercial vehicles ISBN 978-3-8325-3338-0, 2013, 139 S. 34.00 €

8	Jianfei Wang	Thermal Modeling and Management of Multi-Core Processors	
		ISBN 978-3-8325-3699-2, 2014, 144 S.	35.50 €
9	Stefan Simon	Objektorientierte Methoden zum automatisierten Entwurf von modellbasierten Diagnosesystemen	
		ISBN 978-3-8325-3940-5, 2015, 197 S.	36.50 €
10	Sven Reimann	Output-Based Control and Scheduling of Resource-Constrained Processes	
		ISBN 978-3-8325-3980-1, 2015, 145 S.	34.50 €
11	Tim Nagel	Diagnoseorientierte Modellierung und Analyse örtlich verteilter Systeme am Beispiel des pneumatischen Leitungssystems in Nutzfahrzeugen	
		ISBN 978-3-8325-4157-6, 2015, 306 S.	49.50 €
12	Sanad Al-Areqi	Investigation on Robust Codesign Methods for Networked Control Systems	
		ISBN 978-3-8325-4170-5, 2015, 180 S.	36.00 €
13	Fabian Kennel	Beitrag zu iterativ lernenden modellprädiktiven Regelungen	
		ISBN 978-3-8325-4462-1, 2017, 180 S.	37.50 €
14	Yun Wan	A Contribution to Modeling and Control of Modular Multilevel Cascaded Converter (MMCC)	
		ISBN 978-3-8325-4690-8, 2018, 209 S.	37.00 €
15	Felix Berkel	Contributions to Event-triggered and Distributed Model Predictive Control	
		ISBN 978-3-8325-4935-0, 2019, 185 S.	36.00 €
16	Markus Bell	Optimized Operation of Low Voltage Grids using Distributed Control	
		ISBN 978-3-8325-4983-1, 2019, 211 S.	47.50 €